Springer Series in Operations Research and Financial Engineering

Series Editors:
Thomas Mikosch
Sidney I. Resnick
Stephen M. Robinson

For further volumes:
http://www.springer.com/series/3182

Alan J. King · Stein W. Wallace

Modeling with Stochastic Programming

Alan J. King
T.J. Watson Research Center
IBM Corporation
1101 Kitchawan Road
Yorktown Heights, NY 10598
USA

Stein W. Wallace
Department of Management Science
Lancaster University Management School
Lancaster
United Kingdom

ISSN 1431-8598
ISBN 978-1-4899-9212-3   ISBN 978-0-387-87817-1 (eBook)
DOI 10.1007/978-0-387-87817-1
Springer New York Heidelberg Dordrecht London

Mathematics Subject Classification (2010): 90-01, 90C15, 90C31, 90B15

© Springer Science+Business Media New York 2012
Softcover reprint of the hardcover 1st edition 2012
This work is subject to copyright. All rights are reserved by the Publisher, whether the whole or part of
the material is concerned, specifically the rights of translation, reprinting, reuse of illustrations, recitation,
broadcasting, reproduction on microfilms or in any other physical way, and transmission or information
storage and retrieval, electronic adaptation, computer software, or by similar or dissimilar methodology
now known or hereafter developed. Exempted from this legal reservation are brief excerpts in connection
with reviews or scholarly analysis or material supplied specifically for the purpose of being entered and
executed on a computer system, for exclusive use by the purchaser of the work. Duplication of this publication or parts thereof is permitted only under the provisions of the Copyright Law of the Publisher's
location, in its current version, and permission for use must always be obtained from Springer. Permis-
sions for use may be obtained through RightsLink at the Copyright Clearance Center. Violations are liable
to prosecution under the respective Copyright Law.
The use of general descriptive names, registered names, trademarks, service marks, etc. in this publication
does not imply, even in the absence of a specific statement, that such names are exempt from the relevant
protective laws and regulations and therefore free for general use.
While the advice and information in this book are believed to be true and accurate at the date of publication, neither the authors nor the editors nor the publisher can accept any legal responsibility for any errors
or omissions that may be made. The publisher makes no warranty, express or implied, with respect to the
material contained herein.

Printed on acid-free paper

Springer is part of Springer Science+Business Media (www.springer.com)

To our mentors: R. Tyrrell Rockafellar and Roger J-B Wets.

Preface

I do not approve of anything which tampers with natural ignorance.
– *Oscar Wilde*

This book has been in the making since Wallace had a sabbatical in Grenoble, France, in 1996/1997, just after Kall and Wallace [30] was published. Already at that time it was clear that the next step would be modeling. Between then and now several intermediate texts have arisen, all being used for teaching over many years, also by others besides us. At one point several attempts to publish the text were made but failed, possibly because the text was not good enough, certainly because the editors (or their reviewers) did not appreciate that stochastic programming was more than mathematics and algorithms. But maybe that was a good outcome after all. After the two of us teamed up, Wallace with a basic text, King with a publisher and fresh ideas, things have moved forward and all delays have been on us. We don't know how the text will be received, but at least *we* are happy with it. It reflects how we think about our field.

We suppose that our reader faces a decision problem, probably one she is fairly knowledgeable about. But she realizes that uncertainty plays an important role. Very likely, a deterministic model will not do the trick. What issues need to be faced before she has a finished model?

Passing from a deterministic to a stochastic formulation involves many subtleties. In this book, we try to be very detailed about what the issues are: why deterministic models fail to capture stochastics meaningfully, how to think about the information structures, how to rewrite the model so that it becomes consistent with the information structure, how to handle the random variables, how to model risk to find a really good solution.

We believe that in many cases, stochastic programming provides an appropriate modeling framework, in particular two-stage models, where the first stage is to make some investment and the second to operate this investment under uncertainty. On the other hand, we know that for reasonably large cases, and certainly when we face integer decisions, the resulting models are, for many problem classes, not numerically solvable. However, the reasons for using this framework are so compelling, especially in view of the fact that the deterministic approaches do not deliver what they promise, that we need to find a way through this dilemma.

Real and financial options and insurance contracts provide robust decisions by offering flexibility in the future where uncertainty must be faced. But all these instruments (decisions) are of the type "Add something on top of something else," for example adding some capability or capacity on top of the solution to a deterministic program or on top of whatever is being done today. These types of instruments already exist, and the main issue is to value them and determine if they are good deals or not.

We show in this book that stochastic programming models provide good solutions because stochastic programming explicitly models the value of future decisions that are made after uncertainty has been revealed. We shall see that the really good solutions, that take future decisions optimally into account, are not of the type "something on top of something else."

So where does this take us? First, we need to understand *why* deterministic optimization does not bring us where we want to be when modeling uncertainty. Next, we need to be able to formulate what we would like to solve. In some cases, we may even be able formulate and solve exactly what we set out to do; realistic stochastic programs are not always unsolvable. But if not, stochastic programming technologies provide us with tools to explore in what *ways* solutions to stochastic programs are good. Chapter 5 gives a specific example of how good solutions come about by explicitly identifying solutions that optimally open up for future decisions.

Understanding why we need stochastic programs, being able to formulate them, and, finally, finding out what it is that makes solutions good can help us find good solutions without actually solving the stochastic programs. Therefore, this work is much more than a book on how to build unsolvable models. Rather, it shows a way forward so that we can potentially benefit from a modeling framework that, in our view, is the right one for many decisions facing uncertain consequences.

Related Literature

This text fills a hole in the present set of available books. On the one hand there are a few rather technical books on how to solve stochastic programs, such as Kall and Wallace [30], Prékopa [46], Birge and Louveaux [7], and a more recent collection edited by Ruszczyński and Shapiro [48]. On the other hand, a collection edited by Wallace and Ziemba [56] covers a large group of genuine applications and discusses available software. Our new book picks up what is missing: how to take the difficult step from a verbal problem description to a useful mathematical formulation. We do not discuss efficient solution procedures as these are available elsewhere.

Audience

The major target groups are graduate students, researchers (academic or industrial), and university professors interested in modeling decision making under uncertainty. The world is full of uncertainty; what can we do about that when modeling? The book will also be suitable for high-level undergraduates specialized in quantitative modeling.

Required Background

The book assumes the reader already has a basic undergraduate knowledge of linear programming and probability and some introduction to modeling from operations research, management science, or something similar.

Technical Difficulty

The book has two parts. The first part carries all the important ideas and should be read by all readers. This part has a low technical difficulty (apart from some details in Chap. 4), but we intend it to be rather challenging conceptually. This is why we define the audience as we did, despite the fact that, technically speaking, lower-level students could probably access the book. The second part of the book contains examples applying the ideas from the first part. These are not applications in the ordinary sense. The focus is: "Here is a problem, how should we think about it—how shall we model it?" rather than "Here is a successful application of stochastic programming." So the second part tries to show how we would start thinking about modeling the way we suggest in the first part. In the second part, the technical difficulty varies, but the reader can pick and choose examples that she finds interesting, accessible, and useful.

Throughout the book we provide pointers to more advanced literature. On every subject we discuss there is much more to be said. However, we have chosen to avoid most details for several reasons, the main one being simply that we wanted to write a book about the *basics* of modeling with stochastic programming. We believe that there is a more serious need for a book explaining the underlying "whys" than the technically deeper "hows." But also, we try to avoid drowning the conceptual difficulties in technical details. We believe that whenever a student or user has the basic questions and difficulties in place, she can always walk down the alleys of detail reflecting her needs and abilities. Another matter is that a book like this, touching upon so many areas related to stochastic programming, and not only discussing stochastic programming in a limited sense, would indeed be quite a brick, if not a collection of bricks. So we prefer to provide a cornerstone rather than material for a whole building.

The Chapters

1. *Uncertainty in Optimization.* This chapter introduces all the basic ideas
   of the book. It contains our main discussion of why sensitivity analysis in
   deterministic optimization does not deliver what it promises. This is done
   primarily by a simple example. We then relate our work to real option
   theory and discuss in fairly great detail what is meant by robustness and
   flexibility. We also touch upon robust optimization and chance-constrained
   models as alternative approaches. The focus throughout is not on describing the mathematics of these approaches but on discussing what it means
   to use them. What do the methods imply about our understanding of the
   world? We also point out that stochastic programming is primarily about
   transient decision making. Finally, scenario generation gets its first treatment.
2. *Modeling Feasibility and Dynamics.* This is the major theoretical chapter.
   It builds on Chap. 1. Much of the theory is presented using examples, and
   rather than focusing on technical intricacies, we again ask what the issues
   are and what should be done. Much of the focus is on the issue of feasibility
   (should we use constraints or penalties) and information structures. We
   also discuss how to handle tail effects if a problem has too many (often
   infinitely many) time periods.
3. *Modeling the Objective Function.* In this chapter, we discuss what to do
   when a decision leads to an outcome distribution rather than a number. We
   discuss when it is appropriate to maximize expected profit and when risk
   should be considered. Soft and hard constraints are discussed, and penalty
   formulations are compared to option modeling.
4. *Scenario Generation.* When solving stochastic programs, the stochastics
   must be represented by discrete distributions. This is a side effect of the
   chosen solution method and is not caused by the problem under study.
   This chapter follows up on Chap. 1 and explains why a user must be careful what can go wrong if one is not careful and discusses what can be done
   for those not interested in scenario generation as such but still need to
   worry about what they end up doing. We discuss stability questions (relative to scenario-tree generation) and statistical approaches to testing the
   quality of solutions (wherever they come from), and we indicate how the
   two relate. We also discuss in some detail what should be understood by
   a good scenario tree/discretization and how different approaches differ in
   this respect. We go into some detail here as scenario generation is probably
   less settled than all the other issues brought up in the book. At the same
   time, in our view, doing scenario generation correctly is much more important than what the literature indicates, given that our interest is the model
   under investigation and not simply the development of an algorithm. This
   reflects our take on the whole subject of stochastic programming: We are
   thinking in terms of decision making and operations research more than
   mathematical programming as such.

5. *Service Network Design.* This inherently two-stage example comes from logistics and is the summary of a thesis doing exactly what this book advocates, step by step. It should be a very useful chapter for a Ph.D. student wanting to embark on a project of her own. The focus is on modeling, in particular how to create a model that is two-stage from a problem that has infinitely many stages.

6. *Multidimensional Newsboy Problem with Substitution.* This chapter discusses an extension of the famous newsboy model, looking at the case with multiple products, dependent demand, and substitution among the products. The model looks at the demand for high-tech fashion products, and it is of particular interest to see how we can model multimodal demand distributions, which are crucial for this problem.

7. *Stochastic Discount Factors.* This chapter adapts the methodology for modeling stochastic discounting in financial options pricing. It places particular emphasis on deriving discount factors and risk measures from forward-looking data like forecasts and prices of futures and options and shows how to apply the methodology to the formulation of stochastic programming problems. This amounts to modeling in both the primal and the dual spaces simultaneously.

8. *Long Lead-Time Production.* This is the only chapter containing a real example. It sees the problem of long lead times and uncertain demand for both producers and customers. The issue is: How does one model this?

Acknowledgements

We gratefully acknowledge the generous assistance of our supporting institutions: IBM Research, Lancaster University Management School, the Chinese University of Hong Kong, and Molde University College. Robin Lougee and Jenny Lu provided essential encouragement and support. The wonderful quotes on uncertainty were collected by Robin, and Jenny provided nonstop cooking to fuel the final sprint. For this, and for so much more, we give our heartfelt thanks.

Yorktown Heights, New York   Alan J. King
Lancaster, UK   Stein W. Wallace

Contents

1 **Uncertainty in Optimization**   **1**
  1.1   Sensitivity Analysis, Scenarios, What-ifs and Stress Tests   2
  1.2   The News Mix Example   4
     1.2.1   Sensitivity Analysis   5
     1.2.2   Information Stages and Event Trees   6
     1.2.3   A Two-stage Formulation   8
     1.2.4   Thinking About Stages   9
  1.3   Appropriate Use of What-if Analysis   10
     1.3.1   Deterministic Decision Making   11
  1.4   Robustness and Flexibility   12
     1.4.1   Robust or Flexible: A Modeling Choice   12
  1.5   Transient Versus Steady-state Modeling   15
     1.5.1   Inherently Two-stage (Invest-and-use) Models   16
     1.5.2   Inherently Multistage (Operational) Models   16
  1.6   Distributions: Do They Exist and Can We Find Them?   18
     1.6.1   Generating Scenarios   20
     1.6.2   Dependent Random Variables   21
  1.7   Characterizing Some Examples   23
  1.8   Alternative Approaches   24
     1.8.1   Real Options Theory   24
     1.8.2   Chance-constrained Models   27
     1.8.3   Robust Optimization   28
     1.8.4   Stochastic Dynamic Programming   31

2 **Modeling Feasibility and Dynamics**   **33**
  2.1   The Knapsack Problem   33
     2.1.1   Feasibility in the Inherently Two-Stage Knapsack
           Problem   34
     2.1.2   Two-Stage Models   36
     2.1.3   Chance-Constrained Models   37

        2.1.4 Stochastic Robust Formulations 38
        2.1.5 Two Different Multistage Formulations 39
    2.2 Overhaul Project Example 39
        2.2.1 Analysis 41
        2.2.2 A Two-Stage Version 43
        2.2.3 A Different Inherently Two-Stage Formulation 45
        2.2.4 Worst-Case Analysis 46
        2.2.5 A Comparison 47
        2.2.6 Dependent Random Variables 47
        2.2.7 Using Sensitivity Analysis Correctly 49
    2.3 An Inventory Problem 49
        2.3.1 Information Structure 50
        2.3.2 Analysis 53
        2.3.3 Chance-Constrained Formulation 54
        2.3.4 Horizon Effects 55
        2.3.5 Discounting 55
        2.3.6 Dual Equilibrium: Technical Discussion 56
    2.4 Summing Up Feasibility 59

3 Modeling the Objective Function   61
    3.1 Distribution of Outcomes 61
    3.2 The Knapsack Problem, Continued 62
    3.3 Using Expected Values 63
        3.3.1 You Observe the Expected Value 63
        3.3.2 The Company Has Shareholders 64
        3.3.3 The Project Is Small Relative to the Total Wealth
               of a Company or Person 65
    3.4 Penalties, Targets, Shortfall, Options, and Recourse 66
        3.4.1 Penalty Functions 66
        3.4.2 Targets and Shortfall 66
        3.4.3 Options 68
        3.4.4 Recourse 68
        3.4.5 Multiple Outcomes 69
    3.5 Expected Utility 69
        3.5.1 Markowitz Mean-Variance Efficient Frontier 70
    3.6 Extreme Events 73
    3.7 Learning and Luck 76

4 Scenario-Tree Generation: With Michal Kaut   77
    4.1 Creating Scenario Trees 79
        4.1.1 Plain Sampling 79
        4.1.2 Empirical Distribution 81
        4.1.3 What Is a Good Discretization? 81
    4.2 Stability Testing 83
        4.2.1 In-sample Stability 84

4.2.2   Out-of-Sample Stability 85
4.2.3   Bias 86
4.2.4   Example: A Network Design Problem 86
4.2.5   The Relationship Between In- and Out-of-Sample
        Stability 87
4.2.6   Out-of-Sample Stability for Multiperiod Trees 87
4.2.7   Other Approaches to Stability 88
4.3   Statistical Approaches to Solution Quality 88
4.3.1   Testing the Quality of a Solution 88
4.3.2   Solution Techniques Based on the Optimality
        Gap Estimators 90
4.3.3   Relation to the Stability Tests 91
4.4   Property Matching Methods 92
4.4.1   Regression Models 94
4.4.2   The Transformation Model 95
4.4.3   Independent and Uncorrelated Random Variables...... 100
4.4.4   Other Construction Approaches 101
4.5   Choosing an Approach 102

5   Service Network Design: With Arnt-Gunnar Lium
    and Teodor Gabriel Crainic   103
5.1   Cost Structure 103
5.2   Warehouses and Consolidation 104
5.3   Demand and Rejections 104
5.4   How We Started Out 106
5.5   The Stage Structure 107
5.6   A Simple Service Network Design Case 107
5.7   Correlations: Do They Matter? 111
5.7.1   Analyzing the Results.............................. 111
5.7.2   Relation to Options Theory......................... 115
5.7.3   Bidding for a Job................................. 116
5.8   The Implicit Options 116
5.8.1   Reducing Risk Using Consolidation 117
5.8.2   Obtaining Flexibility by Sharing Paths 119
5.8.3   How Correlations Can Affect Schedules 121
5.9   Conclusion 122

6   A Multidimensional Newsboy Problem with Substitution:
    With Hajnalka Vaagen   123
6.1   The Newsboy Problem 123
6.2   Introduction to the Actual Problem...................... 125
6.3   Model Formulation and Parameter Estimation 126
6.3.1   Demand Distributions 126
6.3.2   Estimating Correlation and Substitution 127
6.4   Stochastic Programming Formulation 128

6.5   Test Case and Model Implementation 130
    6.5.1   Test Results 131
6.6   Conclusion 137

7   Stochastic Discount Factors   139
7.1   Financial Market Information 139
    7.1.1   A Simple Options Pricing Example 140
    7.1.2   Stochastic Discount Factors........................144
    7.1.3   Generalizing the Options Pricing Model 144
    7.1.4   Calibration of a Stochastic Discount Factor 146
7.2   Application to the Classical NewsVendor Problem 148
    7.2.1   Calibration of Real Options Models.................150
7.3   Summary Discussion 150

8   Long Lead Time Production: With Aliza Heching   153
8.1   Supplier-Managed Inventory 153
8.2   Supplier-Managed Inventory: Time Stages 154
    8.2.1   Modeling Time Stages 154
8.3   Modeling the SMI Problem 156
    8.3.1   First- and Last-Stage Model 156
    8.3.2   Demand Forecasts and Supply Commitments 157
    8.3.3   Production and Inventory 157
8.4   Capacity Model 158
    8.4.1   Orders and Review Periods 158
    8.4.2   The Model 159
    8.4.3   Objectives.......................................159
8.5   Uncertainty..161
    8.5.1   Uncertain Orders.................................161
    8.5.2   Inaccurate Reporting 162
    8.5.3   A Stochastic Programming Model 162
    8.5.4   Real Options Modeling............................163

References   165

Index   171

Chapter 1

Uncertainty in Optimization

Life is uncertain. Eat dessert first.
– *Ernestine Ulmer*

Decisions are rarely made under certainty. There is almost always something relevant about the future that is not known when important decisions are made.

- What will the weather be during the outdoor concert?
- Will my technology capture the market, or will my competitor capture all the sales?
- Will the government change the taxation rules?
- Will the killer asteroid hit during lunch?
- What will the weather be during the outdoor concert?
- How cold will it be this winter (and hence what will be the energy demand)?
- What will demand be for my product the next 10 years (as I am making a major investment in production equipment)?
- Will the Chinese currency become convertible?
- What is the travel time to my outlet in central London?

There is no end to what we do not know. Still, we must make decisions since the problems are there and demand attention. The key central questions to modeling uncertainty in decision problems will always be:

- What are the important uncertainties?
- How can we handle them?
- Can we deliver valuable solutions and insights?

Stochastic programming is the part of mathematical programming and operations research that studies how to incorporate uncertainty into decision problems. The canonical expression of the decision problems we seek to handle in stochastic programming goes as follows:

A.J. King and S.W. Wallace, *Modeling with Stochastic Programming*,
Springer Series in ORFE, DOI 10.1007/978-0-387-87817-1_1,
© Springer Science+Business Media New York 2012

> Some decisions must be made today, but important information will not be available until after the decision is made.

Problems of this type are found in transportation, logistics, supply chain management, capacity planning, financial asset management, and other related fields. The problems show up in government as well as private enterprise. Stochastic programming provides modeling and solution techniques that harness the power of optimization to solve models that are sensitive to the impact of uncertainty.

> This book is about modeling decision problems in order to obtain solutions that perform well under uncertainty.

This book is also about what the parts of this statement mean: about what a "solution" is and what it means for solutions to "perform well" in applications and environments with much uncertainty.

Before embarking upon this text you have most likely had a modeling course in operations research, management science, or mathematical programming. It is likely that only a minor part of the book you used in the course covered that aspect of decision making. Maybe there was a section on sensitivity analysis in optimization, one on stochastic dynamic programming, one on decision trees, and one on queuing. Did you notice that sensitivity analysis in optimization does not connect very well to the setups for modeling uncertainty? Possibly you did, but the book did not bridge that gap. This book explains what to do next and how to make the connection between optimization and uncertainty.

Current and future operations research professionals such as yourself should learn to analyze uncertainty from the perspective of stochastic programming. Thinking about how to model in an uncertain, dynamic world could be one of your most important contributions to decision making.

1.1 Sensitivity Analysis, Scenarios, What-ifs and Stress Tests

Most, if not all, operations research textbooks advise us to be aware that when we formulate a mathematical programming problem, some of the numbers we need, such as demand, cost, and quality, are not really known to us. This could be either because they are truly unavailable (e.g., the demand for sugar next year) or it is a question of ignorance (e.g., the sales of sugar of different types in Namibia last year). The common practice in such cases is:

- Use statistics, experts, whatever you have to learn about the parameters.
- Solve your model with expected values entered everywhere.
- Observe that your optimal solution $\hat{x}$ depends on the parameters you used.
- Apply *sensitivity analysis* to determine whether $\hat{x}$ depends critically on the parameters.

Sensitivity analysis evaluates the degree to which parameter values can vary without changing the fundamental character of the solution. The general idea is that if you can vary the parameters a lot, your solution is stable relative to variations in the parameters, whereas if you can only vary them a little, you have an unstable solution. The latter case is cause for concern since the parameter estimates may be off and, hence, you may easily be in trouble, whereas in the first case an error in your estimation is not too dangerous. Using the *parametric optimization* tool, you can see how the solution (and the dual variables in the case of linear programs) change as you move away from your initial values. You can then acquire a good feel for which parameters are critical and where you should make an extra effort to obtain better estimates. By this type of analysis, most textbooks claim, you can determine if uncertainty is important or not.

What-if analysis, *scenario analysis*, and *stress-test* approaches all generate a set of possible futures and ask "what if each of these turns out to be the case?" What-if analysis is heavily used in problems involving discrete/integer variables and addresses questions like "what if we open a facility in Hong Kong?" or "what if our competitor opens a distribution center in China?" Scenario analysis and stress tests (in our view) are better applied in situations where there is a need to explore different trendlines in fundamental parameters, such as employment growth, interest rates, or commodity indexes.

All these techniques represent exactly the same kind of approach: to explore a variety of future states of the world and think through the consequences for each one in isolation. Sensitivity analysis is really equivalent in spirit to these what-ifs, scenarios, and stress tests: all are used to generate a set of possible optimal solutions for the next step of the decision process. For most of us, this thinking is very intuitive.

A deeper look suggests that sensitivity analysis approaches are likely to miss important features of the solution. They cannot propose any kind of solution that addresses variability, and they cannot model the natural process of investing in technologies that are useful in responding to uncertain outcomes.

Decisions made on the basis of sensitivity analysis and its relatives can only find solutions that are optimal for some fixed (i.e., deterministic) setting of the uncertain parameters.

These approaches are so common that we feel a need to discuss them very early in this book. Our wish is for you to develop a good understanding of what the combination of a deterministic model with expected values, followed by sensitivity analysis, gives and what it does not give. Our own experience is that whereas many students find it easy to accept that stochastic programming is an appropriate modeling framework when facing uncertainty in a model, it is more difficult to grasp why *sensitivity analysis normally is not a good framework for approaching uncertainty*.

1.2 The News Mix Example

This section contains an example to illustrate important aspects of what can go wrong if sensitivity analysis combined with parametric optimization is applied to a decision-making problem facing uncertainty, followed by a discussion of how to avoid these difficulties.

The example models a variation of the classical news vendor problem, in which the news vendor must make a standing order of newspapers that will be delivered at the start of every day for some period of time:

- Newspaper orders are limited to no more than 1,000 due to the size of the delivery truck.
- The news vendor can choose between three papers: a political newspaper, a business newspaper, or a regional newspaper.
- The profit from the sale of one political newspaper is $1.30, from one business newspaper $1.20, and from one regional newspaper $1.00.
- Survey data show that customers' first choice is either the political paper or the business paper, but their second choice is almost always the regional paper.

We make the modeling assumption that the demand for political and business papers sums to exactly 1,000, though we know that this is not totally true. Let D_p be the demand for the political newspaper. The demand for the business newspaper is then almost certainly $1,000 - D_p$. Table 1.1 summarizes the data for the problem.

Table 1.1: Data for news mix example.

Newspaper	Political	Business	Regional
Quantity ordered	x_p	x_b	x_r
Quantity demanded	D_p	$1,000 - D_p$	∞
Profit	$1.30	$1.20	$1.00

The ∞ used in the demand for regional newspapers simply represents that we can sell whatever we order within our capacity.

We start by formulating the news mix choices as a deterministic optimization model. For convenience, we normalize the numbers by dividing through by the maximum order. Orders are ratios of the maximum order (so they are between 0 and 1), and D_p is the ratio of demand for political newspapers.

$$\max_{x_p, x_b, x_r} \quad 1.30x_p + 1.20x_b + 1.00x_r$$

$$\text{such that} \quad \begin{cases} x_p & \leq D_p, \\ x_b & \leq 1 - D_p, \\ x_p + x_b + x_r \leq 1, \\ x_p, \quad x_b, \quad x_r \geq 0. \end{cases} \quad (1.1)$$

In model (1.1) the news vendor maximizes the "profit" (objective function) over all orders $[x_p, x_b, x_r]$ that do not exceed demand, sum to no more than one, and are nonnegative.

1.2.1 Sensitivity Analysis

Since the ratio D_p is not known, a reasonable initial approach might be to solve model (1.1) for *any* possible value of D_p. The optimal solutions are in this case easy to find and are given by

$$\begin{aligned} x_p &= D_p, \\ x_b &= 1 - D_p, \\ x_r &= 0. \end{aligned} \quad (1.2)$$

Suppose now that the forecast for the ratio of political to business newspaper demand is $\bar{D}_p$. Then by (1.2) we can find the optimal order quantities for this estimate demand:

$$\begin{aligned} \bar{x}_p &= \bar{D}_p, \\ \bar{x}_b &= 1 - \bar{D}_p, \\ \bar{x}_r &= 0. \end{aligned}$$

We will put this solution into the news mix model (1.1) and see what happens if we one day observe a demand ratio $\hat{D}_p$. The constraints become

$$\begin{aligned} \bar{D}_p & \leq \hat{D}_p, \\ 1 - \bar{D}_p & \leq 1 - \hat{D}_p, \\ \bar{D}_p + 1 - \bar{D}_p + 0 & \leq 1. \end{aligned}$$

Consider the first two. One of them must be violated for any $\hat{D}_p$ that is not precisely equal to $\bar{D}_p$. Unless the demand turns out to be *exactly* $\hat{D}_p = \bar{D}_p$, the model will be infeasible!

What possible conclusion can we draw when our model is practically always infeasible? Does "infeasibility" even make sense in this situation?

Well, it probably does not. The fact is that the model we started with is just badly suited for analyzing the problem for unanticipated data values.

> An often overlooked step in modeling is the determination of how to handle *unanticipated values* of the model data.

You may think this is obvious. Surely you can think of several better models than (1.1) for the newsvendor. But what if the problem was really, really complex with lots and lots of parameters and constraints? Would you be able to tell that the problem was badly formulated? Parameter estimates are *always wrong*. They are just estimates, after all. The difficulties we just observed in our little problem may be hidden within a large model, and it will not be possible to directly observe what is going on.

> Do not trust that you can *see* whether or not a large model is suited for parametric analysis. You need to prepare for it by proper modeling.

When the modeling does not tell you what to do when parameters turn out to be wrong, then it is likely that serious issues will emerge only when investments have been made and plans put into practice. Deterministic models, by definition, do not say anything about what to do when parameters are not as expected. For stochastic programs, on the other hand, modeling what might happen and how to handle each and every situation is the core of the model.

The property we have just observed is called a *knife edge*. Optimization procedures tend to do anything to save even a cent, and this results in solutions that fit perfectly for the given parameters but can be very bad for even very similar values: They balance on a knife edge. We have just modeled an extreme case of this for the news vendor, namely, a solution that is optimal for one parameter value but infeasible for all others! This was to a large extent caused by weak modeling. But even in more well-behaved cases, we may observe these knife-edge properties. To some extent, this property is the major reason why solutions from deterministic models can have very bad average behavior.

1.2.2 Information Stages and Event Trees

The difficulty we faced when analyzing the news mix formulation could have been addressed if we had equipped model (1.1) with *stages*.

An information stage (normally simply called "stage") is the most important new concept that distinguishes stochastic programming. A stage is a point in time where decisions are made within a model. Stages sometimes follow naturally from the problem setting and sometimes are modeling approximations.

It only makes sense to distinguish two points in time as different
stages if we observe something relevant in between.

"Observing something relevant" must be widely understood. If we hope to
receive some data before Tuesday morning but they do not arrive, then that
is, of course, information: we now know the data did not arrive, and so we
must do without them.

The importance of stages was not so easy to see in the deterministic version
of the news mix model (1.1). However, when we implemented the particular
solution $\bar{x}$ and saw what happened when demands became known on any given
day, then it became clear that there were in fact *two stages*, as illustrated by
the event (or scenario) tree in Fig. 1.1.

In *stage 0*, orders are placed. In *stage 1*, after the demand for newspapers
has become known, we determine what can be sold. With the two stages in
place, it seems rather clear that letting the same variables represent both
orders and sales is not a very good idea. Because that is what happened: the
variables were interpreted as order ratios, but some of the constraints referred
to sales. In a deterministic setting, you will never order something that cannot
be sold, so using the same variables for orders and sales makes some kind of
sense.

In a deterministic world you will never buy or produce something
you will never need. This often leads to models with structures that
do not even allow meaningful sensitivity analysis.

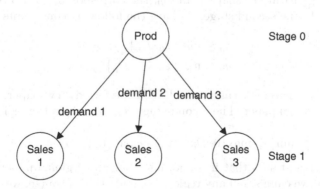

Fig. 1.1: Event (or scenario) tree describing a stage structure for news mix
        example

We have already noted that (1.1) is reasonably well defined in a deterministic environment. We have also noted that it produces rather strange

conclusions when analyzed in light of the uncertain demand. One reason for this is that the stage structure is not properly reflected in the formulation.

Let us review the situation. We will receive demand for political and business newspapers that is close to 100% of our maximal total order, but at order time, the split between political and business newspapers is unknown.

What happens if we choose one of the solutions from the deterministic model (1.1), say $\bar{x}_p = \bar{D}_p$, $\bar{x}_b = 1 - \bar{D}_p$, and $\bar{x}_r = 0$, but it turns out that the actual ratio does not equal the estimate $\bar{D}_p$? Will the world end? Of course not. Will the news vendor go broke? Probably not. Will nothing be sold since we got it wrong? Unlikely. But the interpretation, within the model, of something being "infeasible" is exactly that: cannot be allowed to happen under any circumstance.

If we get the estimates wrong, it is more likely that we will face a penalty, either directly or indirectly in terms of reduced sales. Now, what happens if we order the wrong amounts (which we almost certainly do)? The fact that the order will be wrong must be incorporated into our model for it to be useful. As you have noted, the stochastic setting with stages has already forced us to ask a lot of questions that are not needed in a deterministic setting, and there are more to come.

1.2.3 A Two-stage Formulation

Let us think about our model in two stages. First we order x_p, x_b, and x_r, constrained by a sum of one. Then demand becomes known, that is, the customers show up, and we learn the values of demand ratios for political papers D_p and business papers $1 - D_p$ (remember that we have assumed that these ratios sum to one). How much can we now sell? Let y_p be the ratio of papers sold as political newspapers and y_b the corresponding ratio of business newspapers. In the second stage, we have the following constraints on y:

$$y_p \leq \min\{x_p, D_p\},$$
$$y_b \leq \min\{x_b, 1 - D_p\}.$$

When our customers cannot find their preferred newspaper, they will choose the regional paper. The second-stage constraints on the regional paper are

$$y_r \leq \min\{x_r, \max\{0, D_p - x_p\} + \max\{0, 1 - D_p - x_b\}\}.$$

This constraint says that if we do not order any regional papers (that is, $x_r = 0$), then we cannot sell any regional papers. But of course, we know that we will order the wrong amounts of political and business papers. So one of the terms in the max expression will be positive. So either we will not have enough political papers ($D_p - x_p > 0$) or we will not have enough business papers ($1 - D_p - x_b > 0$). Either way there will be a demand for regional papers!

Thinking in two stages has revealed an important aspect of the news mix problem—that the demand for regional newspapers is always going to be nonzero, caused by customers' not finding their first-choice paper. This is something we see when we analyze this problem in terms of stages. Now, perhaps a good business person does not need a course in stochastic programming to see this point. She would always order a few copies of the regional just because it is not good business to turn away customers!

But do you see that there is something unusual going on? Notice that there is no demand scenario under which people will order a regional paper. They always prefer either the political or the business paper. But the news vendor cannot know in advance what this demand will be, and so in order not to turn away too many customers, the vendor will order some regional papers. In a sense, the news vendor is buying a hedge against the uncertain demand for political and business papers. This is worth an observation:

> Even if some property is present in the optimal solution to *every* deterministic version of a problem, this property does not necessarily carry over to the optimal solution when uncertainty is taken into account.

In fact, the chance that such a property will carry over is rather slim. This is crucial to observe: There are aspects of a good solution that deterministic models will never reveal, no matter how much you analyze them.

1.2.4 Thinking About Stages

A useful model needs to capture the salient features of a decision problem. The analysis required to model the sequence of interrelated decisions includes much more than just estimating distributions in situations where the deterministic model would require numbers. New questions arise. We have to *identify stages* and *define variables consistent with the stage structures.* We believe that a good modeler should think carefully about stages independently of what kind of model is ultimately used. Stochastic thinking simply helps in understanding the decision problem better, or it at least helps to shed light on how little we have understood.

Issues related to feasibility showed up in our example in questions like "what if we have ordered too little?" or "what if we have political newspapers left but unsatisfied demand for business newspapers?" We also note that there are numerous stochastic versions of one deterministic model. For many of the questions we asked—and answered—there are other potentially correct answers, and they would have led to different stochastic models.

A useful approach is always to consider the time sequence in which decisions are made and the amount of information available at each point in time. If this exercise identifies important relationships between the decisions, it may be necessary to incorporate stages in the model. This is true whether you end up with a stochastic or a deterministic model.

1.3 Appropriate Use of What-if Analysis

There is one case where what-if analysis will yield the correct answer, and that is when the full first-stage decision is the same for all values of the random variables. If what we do under full information is not changed by that information, the suggested solution will *always be optimal in hindsight*. In this situation, flexibility with respect to the future is unnecessary.

> If what-if analysis (or parametric optimization) leads to the same full first-stage decision for all possible values of the random variables, then this is indeed the optimal solution to the corresponding stochastic programming model as well.

It may be useful to consider a case, for example, where the profit of an operation is dependent on a random variable but the optimal decision is not. Suppose the random variable is demand. If demand is high, you get rich; if demand is low, you barely get by. But whatever happens, the optimal choice is always to build a plant of a given fixed size. The stochasticity is important, but you cannot do anything about it. In such a case, a deterministic model will give the same solution as a stochastic model.

Do not confuse this case with a situation where a certain aspect of the solution is present in all optimal solutions of *deterministic* versions of a problem. In the news mix problem, the optimal solution never contained regional newspapers, despite the fact that it is clearly optimal to order some of them. It is only when the *whole* solution is unaffected by uncertainty that you can draw the conclusion that uncertainty does not matter.

> If what-if analysis (or parametric optimization) leads to a certain aspect of the solution being the same for all possible values of the random variables, then there is no reason to conclude that this aspect will also be part of the optimal solution to the corresponding stochastic programming model.

If a certain aspect of a solution shows up in all what-if solutions, you may be facing something that needs to be done in any case (like establishing a certain flight for an airline since without it the whole network would make no sense), or you may see the effect of ignoring stages (like ordering no regional newspapers in our news vendor model). Distinguishing one from the other may not be easy.

1.3.1 Deterministic Decision Making

Apart from the special situation just discussed, the general observation is as follows:

> Sensitivity analysis is an appropriate tool for analyzing *deterministic* decision problems.

Maybe it is not so surprising that a deterministic tool functions best for deterministic models. The additional complexity introduced by uncertainty in stochastic models makes them more difficult to analyze. New techniques are needed. Simple tools are often used out of expedience, but the usefulness of the results is lacking. There are better alternatives. Using simple tools will not make the complexity of a model go away.

So sensitivity analysis will work in the following situation. Assume you will be making a decision next year. When the time comes, everything will be known. But presently many parameters of the problem are uncertain. You will need to ask yourself: What will I be deciding? If you need to create a budget for next year, what are reasonable estimates for costs or profits? This is a deterministic problem analyzed under uncertainty, and a deterministic model combined with sensitivity analysis would be an appropriate way to approach the problem. The reason is that no decisions are made under uncertainty. The problem has only one stage, and hence a model with one stage (a deterministic model) will be the right choice.

Note that we sometimes use what-if questions to represent decisions: "What if we buy our competitor?" "What if we outsource our warehouses?". This has nothing to do with uncertainty. It is simply a way of expressing a possible decision, and there is nothing wrong in stating a problem in such a fashion. But if we ask: "What if our competitor outsources her warehouses?". Because if this would seriously affect our business and we are uncertain as to what she will do, then using a "what-if" approach (as discussed in this chapter) is not wise. We need to take into account the uncertainty and come up with a decision in light of the uncertainty. It is important to see the difference between these two cases.

In engineering we often find deterministic models of the worst-case type. A deterministic model is set up to make sure that a bridge, for example, can support a load of at least a certain weight. We should not mistake this deterministic model for a deterministic problem. The bridge most likely can support a load different from the one specified (probably higher), but there is also a chance that it will be subjected to forces not previously observed. Winds of unusual strength and direction not previously seen may occur, or the bridge may be hit by a ship, or there may be a fire that affects the load-bearing properties of the steel. Statements about the performance under future loads will always be probabilistic. Deterministic modeling does not change the stochastic nature of a problem, and statements of the type will (always) are (almost) never true.

1.4 Robustness and Flexibility

> I can't change the direction of the wind, but I can adjust my sails to always reach my destination.
> – *Jimmy Dean*

When making decisions under uncertainty, we may look for decisions that help us to (1) withstand random events or (2) accommodate those events. We define these decisions as robust and flexible, respectively. To picture the situation, a tree is robust with respect to heavy winds, while grass is flexible. Some define flexibility as the ability to bring a system from a disturbed state back to where it "should" be. A close look will show that this definition is the same as ours.

1.4.1 Robust or Flexible: A Modeling Choice

A decision does not happen to be robust or flexible. These are qualitative properties we force upon them via a model. In any solution to a model, there will be some decision variables that are robust relative to a random phenomenon, others that are flexible. If there are several random phenomena, as there normally are, a variable may be robust relative to one of them and flexible relative to another. The choice simply depends on whether the decision is made before or after the realization of the corresponding random variables.

To illustrate the robust-versus-flexible issues, let us introduce a pair of what we will come to call "inherently two-stage" problems.

- *Truck routing.* You are setting up routes for trucks in a trucking company. It is part of the agreement with the labor unions that the routes will be kept unchanged for 6 months. But the demand is both varying and uncertain. How do you set up the routes for the next 6 months when you take into account that you will be able to send the goods with any of your trucks as long as the goods get to their destinations on time.

  If we set up a two-stage model, where the first stage determines routes for the trucks (in light of random demand) and the second stage, after demand has become known, determines how to dispatch the goods on the trucks, we have *modeled* a situation where truck routes are robust and dispatching is flexible relative to the random demand. If we also allow partial rerouting of trucks (at a cost) in the second stage, then truck routes are also (partly) flexible. This example illustrates a general principle: In a two-stage model, robustness is in the first stage (truck routes) and flexibility (dispatching) is in the second stage. Notice that the terms flexible and robust are purely qualitative. Being robust does not mean being good or bad; it just means that uncertainty is met by doing the same thing no matter what happens. The trucks have schedules that they follow irrespective of realized demand: the truck schedules are robust. Since we allow dispatching to use knowledge of demand, dispatching is flexible with respect to random demand.

- *Sports event.* You regularly organize a large sports event. But you are troubled by bad weather. If it rains, few people show up. How can you protect yourself against the negative effects of the weather?
  Let us assume you want to maximize expected profit. What can you do? There are some first-stage decisions you can make. One might be to build a roof on your stadium. The roof obviously provides robustness relative to the random weather. You may also pay an insurance premium against bad weather. The insurance provides robustness against bad weather, not by making the weather irrelevant for the sports event, but by providing flexibility in how you obtain income: instead of running the event, you can collect on the insurance. You may also be able to pay a fee for the right to use another stadium in the case of bad weather. The right provides robustness against bad weather by providing flexibility in the operational phase: you either use your own stadium or, in the case of bad weather, the alternative one.

The quality of a solution is defined relative to the objective function. But robustness and flexibility are defined relative to how the actions represented by the variables are set in time as compared to the realizations of the random variables. This is not how all people use the terms. Many use *robust* to mean a solution where the objective function is stable (does not vary too much) over the realizations of the random variable, e.g., "This plan is robust as my income is almost the same whatever happens. I am protected against uncertainty." Some will use only variables that are robust, in our terminology, to achieve this stability in income; others will also use flexibility, depending on their modeling philosophy. Our view is that robustness and stability are qualitative properties of decisions and represent different strategies for facing uncertainty. We reserve the terms *good* and *better* to refer to properties of the objective function.

If stability is required, it will be expressed in the objective function or as a constraint, in which case *good* will be a property of decisions that gives stability. The reason for this is maybe a bit subtle: while there is always a need to face uncertainty, stability in the objective function is often not a goal. So we will make statements like "These truck schedules are robust relative to the random demand, the routing of goods is flexible, and the overall expected profit is maximized." Or "The truck schedules are robust, but with some minor flexibility in changing them after demand is realized, routing of goods is flexible, and, as required, the income stream is stable." We do not want *robust* to be reserved for the special case of income stability, which often is not required.

Assume you have some initial solution that may come from a model or simply may be the normal way of doing business. We may then ask: Are there any (real) options available for which the initial deterministic cost is smaller than the impact of the option on our uncertain outcomes? If the answer is yes, you buy the option, if not, you do not. Notice that buying an option is a robust way of facing uncertainty as it takes place before you observe the outcome of the random variable. The option provides flexibility in a later stage. This connection between robust and flexible is typical and is best seen in the light of options:

> Buying (or selling) an option is a robust way of facing uncertainty, and it has value because it increases flexibility in a later stage.

1.4.1.1 Robust or Flexible?

In some cases, we wish to have robustness. In other cases, we want flexibility. In some cases, either one will do. A bus schedule, for example, should be robust, as passengers normally will hate repeated changes in the schedule. The published schedules (i.e., the services offered) in a company like DHL or FedEx should also be robust, whereas we would not mind it if the routing decisions were flexible, meaning that goods could be sent along many different routes. In a production process, we might find flexibility and robustness equally interesting if the impact are the same.

Although far from being the main reason for choosing between a robust and a flexible approach to handling a random phenomenon, you should be aware that there is a connection between this choice and the question of bounding a decision problem under uncertainty. This way of thinking can sometimes be very useful. Let us illustrate this by two examples.

1.4.1.2 Flexibility Bound

Options pricing in mathematical finance assumes that the instrument can be traded all the time, that is, continuously *without any transaction costs*. This is a perfectly flexible decision setting, so this comprises an optimistic view of the trading capabilities of the decision maker. For a seller, this will be an upper bound, for a buyer, a lower bound. If you choose to model something as being more flexible than it really is, then you obtain a bound. Sometimes this is a wise thing to do, not because you want a bound, but because the actual situation is simply too complicated. One nice effect of thinking like this is that you may be able to verify that you have a bound rather than just an approximation.

1.4.1.3 Robustness Bound

In a robust approach to decision making, models have only one stage. There is no modeling of what to do after we have learned what has happened. If there exist ways to react to uncertainty after uncertainty has been revealed, then the robust approach will be pessimistic. For example, if you are setting up truck schedules for a transportation company, and you assume that, no matter what happens, you cannot change the schedule, then you have made all scheduling variables robust. If in reality you could change the schedule to some extent after you know what has happened, then your robust approach is a bound.

> You can use the choice between robustness and flexibility to bound a problem.

1.5 Transient Versus Steady-state Modeling

In modeling, we often distinguish between transient and steady-state models. A steady-state model is one that, tells us what to do for every state that can possibly occur. There might be states that are only reachable if we do not make optimal decisions, but these are often excluded in calculations. Transient models, on the other hand, start out observing where we are and then tell us what to do. That is, we are somewhere, we have a goal, some resources, and some constraints, and we try to do our best.

> Stochastic programs are primarily about transient decision making.

The starting point of a stochastic program is the present situation. The model is intended to tell us what to do in light of our goals, constraints, and resources. We might, for example, be running a transportation company. We have a fleet and some customers, must obey some rules, and have a goal for our company. We then ask what we should do. We are not asking what to do in all possible situations, just what to do based on our present state.

If there is no memory in a model, then there is no essential difference between steady-state and transient modeling. But if there is memory, which is normally the case for important decisions, stochastic programming is primarily about transient modeling. We are where we are, be that good or bad, and we want to know what to do. Even optimal decisions might leave us worse off than we had been before we made the decisions if constraints are tough and resources limited.

1.5.1 Inherently Two-stage (Invest-and-use) Models

An inherently two-stage model is a model where the first decision is a major long-term decision, whereas the remaining stages represent the use of this investment. This could be something like building a factory for later production under uncertain demand, prices, or even products. Or it could be to set schedules for a certain period of time for an airline, schedules that will face uncertain demand and flying conditions as well as technical problems on planes. In these cases, mathematically speaking, the first stage will look totally different from the remaining stages, which will all look more or less the same. This is a very important class of models, possibly the most important one for stochastic programming.

1.5.2 Inherently Multistage (Operational) Models

In inherently multistage models, all stages are of the same type. Typical models come from financial trading, production/inventory modeling, or some vehicle routing models. Also, here we face things such as random demand, prices, or even products, but the models are not split into an investment part and a usage part. We should not confuse information stages with time periods. Stages model the flow of information; time periods represent the ticking of the clock in a model. Stages, on the other hand, are points in time where we make decisions in the model after having learned something new.

1.5.2.1 Electricity Production Model

To illustrate the basic issues in modeling stages and time periods, we introduce the problem of electricity production

- *Electricity production.* You are responsible for production planning for a major utility company. You face uncertain demand, uncertain rainfall for your hydro-based production, and uncertain wind for your windmills. How do you plan your operations for the next day?

Assume we model daily electricity demand and production over a week. The model has seven time periods. We could create a model with an equal number of stages, which would require, for example, that we model the following sequence (where we assume that decisions for a given day are made *before* demand for that day is revealed; see also Fig. 1.2):

- Decisions are made Monday morning for Monday production.
- Demand information for Monday arrives.
- Decisions are made Tuesday morning for Tuesday production.
- Demand information for Tuesday arrives.
- . . .
- Demand information for Saturday arrives.

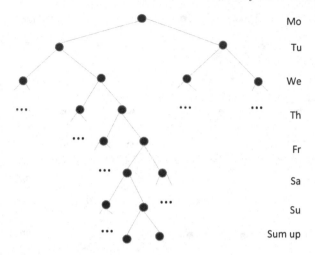

Fig. 1.2: Stage structure when we have one stage per day. For simplicity
we use only two outcomes of the random variable per day

- Decisions are made Sunday morning for Sunday production.
- Demand information for Sunday arrives.

Modeled this way, the model has as many time periods as it has stages. But
we might find that this is a bit detailed. We might realize that the model will
be rerun every day, so in fact, only the first set of decisions will be implemented
(in our example, the Monday decisions). Thus we instead create the following
model (also illustrated in Fig. 1.3):

- Decisions are made Monday morning for Monday's production.
- Demand information for Monday arrives.
- Decisions are made Tuesday morning for Tuesday's and Wednesday's pro-
  duction.
- Demand information for Tuesday and Wednesday arrives.
- Decisions are made Thursday morning for production the rest of the week.
- Demand information for Thursday to Sunday arrives.

This latter model still has seven time periods, but only three stages: Monday, Tuesday, and Thursday mornings (it is a question of taste if you also
count the final "Summing up" as a stage). So we see how the arrival of information has been aggregated and the number of stages reduced. Finding a
good trade-off between time periods and stages is often crucial when modeling
as it has consequences for model quality, data collection, and solvability of the
model.

Inherently multistage models can become inherently two-stage models if
we add a stage with a qualitatively different decision framework. For example,
by adding an initial investment (buying vehicles, building warehouses or
factories), you end up with an inherently two-stage model with the first stage

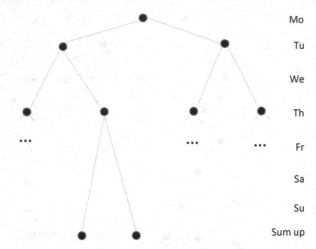

Mo

Tu

We

Th

Fr

Sa

Su

Sum up

Fig. 1.3: Stage structure when we have fewer stages than time periods. For simplicity we use only two outcomes of the random variable per stage

as an investment stage and the remaining periods forming a second stage that provides information about operational costs and benefits to the initial investment. Chapter 2 discusses a technique for modeling the *end-of-horizon* issue in multiperiod decision problems; we will also see there that many operational stages become embedded into a single stage. It is important to note that the inherent structure is a modeling statement, unrelated to the actual number of stages (decision points) in the model.

Multistage models are, in our opinion, less suited for stochastic programming than their inherently two-stage counterparts. This is particularly true if the models are set in a kind of steady-state environment. Stochastic programs focus on the transient aspect of decision making. A financial investor, for example, may find herself stuck with bad investments, high transaction costs, or violations of financial regulations. In this case, a stochastic program might be appropriate. But an investor who finds herself well positioned might be better served by a steady-state model.

1.6 Distributions: Do They Exist and Can We Find Them?

It is tempting to reply "no" and then "yes" on these questions. That obviously requires some explanation. In our view, we must approach this issue with care. On the one hand, there is the distinction between the existence of a distribution and our ability to find out what it is. On the other hand, there is the distinction between repeated and one-time events. And there are also cases where the situation we face will become a repeated event, but so far we have no experience with the case. And we should never forget that we are *modeling*, that is, we are attempting to simplify the real world, so as to learn about it. That obviously also implies modeling, and hence simplifying, the description of random phenomena.

Whatever view we have on how random phenomena should be approached, in stochastic programming, we are always talking about the future. Apart from such things as odds at a casino or the probability of winning the lottery, it is an *assumption* that whatever we might know about the past will be relevant for the future. So even if we have a frequentistic view of life, and even if we have loads of data, how do we know that the future will have anything do to with the past? Although the authors of this book believe we can say something useful about the future, and that mathematical modeling is a useful tool, we should never forget that this is an assumption or a belief. We can certainly test, in most cases, in the past we made good descriptions of the future, but that is as far as we can take it. The assumption that this is still true cannot be tested.

You may say that we have the whole machinery of statistics to help us state something about the future. But using statistics this way exactly implies the assumption that the past does describe the future. There have been a fair number of financial crises over the last few years showing that at critical points in time, the past does *not* describe the future well.

> We can never test if our description of future events is useful or good. This remains an assumption or a belief.

The view we take here is in some sense more pessimistic than the view of many who object to stochastic programming due to the need for distributions. However, the issue is not how optimistic or pessimistic we are, but how we move from this point.

As we have seen in this chapter, sensitivity analysis, what-if analysis, and the like may not deliver what they promise. We miss out on important aspects of robustness and flexibility in the solutions. If you face a problem with an important random variable, even if you know very little about the actual distribution, we would strongly advise setting up a stochastic program with two or three scenarios (three may be wise so as to obtain mass in the middle). That way your solution will change dramatically in structure, and you will learn a lot about your problem. This might mean solving a problem about twice the size of the original problem, something you can probably handle. Not to do this because you do not know the distribution is, in our view, a very bad choice. Chances are that by using a distribution you will obtain a reasonable feeling about the medicine, if not the dosage.

It is a peculiar argument that uses the lack of precise knowledge of distributions as an argument for using deterministic modeling. Putting all the probability mass in the mean of the distribution is certainly neither right nor good.

Alternatively, many prefer to use worst-case analysis to analyze a problem. They may work with intervals and come up with results that do not depend on the actual distributions. This might sound tempting, but in our view it is often not good modeling. Worst-case analysis over an interval will normally amount to putting all the mass in the endpoints of the interval, which we find to be a very peculiar distribution. The results might be very sensitive to the choice of interval, and that is not good if the interval is subjectively chosen.

Supports of random variables in stochastic programming are also most of the time subjective. But you are aware of this issue, of course! The use of low probabilities in the tails offsets this to some extent. Soft rather than hard constraints (that is, we put the constraints into the objective function with a penalty for violation) makes tail effects much less serious.

> Distributions, combined with soft constraints, should be used to learn about the decision problem.

Information we might have about a phenomenon, even if that information is subjective or circumstantial, should be used. If we have good data, we should use them. If we have expert opinions, we should use them. If all we have is our own humble understanding of the problem, then we should still create distributions so as to obtain a feeling for what constitutes good solutions to our problem. We should be able to find the right medicine, even if the dosage is off. After all, we are modeling, and we seek knowledge and understanding.

1.6.1 Generating Scenarios

The generation of scenarios from distributions (wherever they might come from) requires some care. This might seem to be in contradiction to our very subjective, loose view on distributions expressed previously. While we view the creation of a distribution as a modeling issue, we still need to be sure that what we solve is what we think we solve. Otherwise our understanding from the modeling may be driven by errors in the scenario generation.

This is why, in a book on modeling, we present a full chapter on scenario generation. Even if you are not at all interested in the subject as such, you should be concerned about the relationship between your model, particularly the description of uncertainty, and the stochastic program you solved. If your interest is the model and real decision making and not merely the ability to solve stochastic programs, you should be concerned that the solutions you are studying (and possibly implementing!) are driven not by how you make scenarios, but by the actual problem formulation.

1.6.2 Dependent Random Variables

If you replace all random variables by their means, you end up with a deterministic optimization problem. Much of this book is about why that might not be a good idea, especially when there are opportunities to learn about the information and take action. A related sense in which expected values are misleading is the fact that they obscure *dependencies* among these random variables. There is nothing in a deterministic model that even tries to represent dependencies.

In addition, you will often see in the literature on stochastic modeling, especially in steady-state modeling, that random variables are assumed to be independent. Stochastic programming has the tools to be able to come to grips with dependencies in an explicit way.

1.6.2.1 Risk Management

Dependencies arise in risk management in a big way. The canonical example is as follows.

- *Risk Management.* You are responsible for managing risk in a large company, with several divisions and lines of business. They sell different products and services whose profitability depends on many different factors. How do you go about managing the overall risk, rather than the individual risks?

Let us look at a very simple example. Assume we have a company consisting of two divisions, each facing the same distribution of profit in the next period:

Profit	−1	0	2	4	5
Probability	0.1	0.2	0.4	0.2	0.1

The expected value is 2, which is considered OK. But there is an understanding that the negative profit is problematic, so each division decides to buy insurance against it. If the profit is −1, which happens with 10% probability, the divisions will receive 1 from the insurance company. The expected return on the policy is 0.1, for which they will have to pay 50% extra, that is, 0.15. The profit distribution (including the insurance premia and payouts) becomes:

Profit	−0.15	1.85	3.85	4.85
Probability	0.3	0.4	0.2	0.1

The mean is now 1.95, a reduction by 0.05, but that gives us an acceptable risk profile.

At the company level, the mean has been reduced from 4 to 3.9, attaining the following profit distribution:

Profit	−0.3 3.7 7.7 9.7
Probability	0.3 0.4 0.2 0.1

So it does not matter if each division insures its risk or the parent company insures both divisions at the same time.

Is this right? Think about it before you proceed.

Did you realize that our assumption above was that the two profit streams were perfectly correlated, in particular, that the bad cases would occur together? Is that reasonable? It might be, of course, but such perfect correlation is rare. What if we instead assumed that they were uncorrelated? It takes a bit of work, but you should be able to calculate it easily enough as follows:

Profit	−2 −1 0 1 2 3 4 5 6 7 8 9 10
Percent	1 4 4 8 16 4 26 4 16 8 4 2 1

But this is less risky! The company can lose 2 with a probability of 0.01 and 1 with a probability of 0.04. So a policy bringing this up to zero would have an expected payoff of 0.06. If there is still a premium of 50%, the cost will be merely 0.09 rather than 0.30, but more importantly, the loss in expectation will have been reduced from 0.1 to 0.03.

But what if the profits are perfectly negatively correlated? In this case there is no risk at all to the parent company! So if the divisions made their own risk adjustments, there would be a certain loss of 0.1, with no benefit at all.

> Correlations and more complicated forms of covariation are crucial to capture when planning. Otherwise, all measures of risk can be totally off.

What do we learn apart from the fact that correlations are important? We learn that it is very dangerous to look at an activity part by part. What looks risky may not be risky at all! Almost all projects and activities are part of a larger portfolio. Risk must be measured at the portfolio level. Even projects with negative expectations may be profitable for a company! (Why?)

It is our view that much of the literature treating risk fails to capture the portfolio effect. The net result is a poor understanding of risk. Stochastic programming is well set up to model any kind of covariation. This is contrary to many other tools (but not contrary to all tools, of course). We will discuss this to some extent in Chap. 3.

1.7 Characterizing Some Examples

Let us look at four typical decision problems under uncertainty that we have seen so far to analyze their main characteristics. These will be discussed in more detail below, but you should think about how you might model using the principles we have presented in this chapter.

Truck routing on p. 12—to be discussed in more detail in Chap. 5

Stage structure: This problem is inherently two-stage with a finite (but large) number of stages as the plan will last for 6 months.

Random variables: Demand for the different commodities is assumed to be random.

Objective function: The goal here is to minimize expected cost. It is reasonable to assume that the average behavior of the problem will be observed and that we therefore are not risk averse.

Why not deterministic? A deterministic model would produce routes that are not very flexible with respect to variation in demand. We will see that in Chap. 5.

Sports event on p. 13

Stage structure: This problem is inherently two-stage with, principally, an infinite number of stages as you have no plans to stop this activity.

Random variables: The weather, particularly the probability of rain.

Objective function: Most likely you will minimize expected costs as you will observe the average performance of your business. But if each event is very large and the events are far apart in time (even if regular), you would likely want to model some risk aversion.

Why not deterministic? In a deterministic model, it would either not rain (in which case the optimal solution is to do nothing) or rain (in which case you would choose the cheapest way to get the event under roof or simply give up.) Other solutions that trade off the two situations (and possibly take into account weather forecasts) would not be valued.

Electricity production on p. 16

Stage structure: This problem is inherently multistage with, principally, an infinite number of stages as you have no plans to stop this activity.

Random variables: Both supply and demand are uncertain, being mostly driven by the weather.

Objective function: You would likely minimize expected costs here as these are repeated operations where you will observe the average behavior.

Why not deterministic? Deterministic models will either produce solutions that cannot handle noise when it occurs or that end up overly pessimistic (and hence costly). A major issue is that many production units have longer startup/closedown times than the time interval over which wind can be predicted.

Risk management on p. 21—to be discussed in more detail in Chap. 3

Stage structure: This problem is inherently multistage with, principally, an infinite number of stages as you have no plans to stop this activity. If you were considering some major changes in procedures or some major investment that would change the whole portfolio, this problem might be inherently two-stage.

Random variables: Market developments and success of individual projects.

Objective function: Here you wish to model risk aversion probably caused by some of the projects having correlated profits. If all divisions and projects had uncorrelated profits, there would be no risk management problem to attend to.

Why not deterministic? Risk management makes little sense in a deterministic world.

1.8 Alternative Approaches

The rest of this chapter will be used to outline other approaches to decision making under uncertainty. Our goal is to put the different ideas into perspective and see where they agree and where they differ. We would also like to point out the strengths and weaknesses of each. We will only present basic ideas and point to principal difficulties, modelingwise, from using the ideas.

1.8.1 Real Options Theory

This theory comes from economics rather than operations research, which is the home of the other methods we cover in this book. The starting point in options theory is "what is it worth" rather than "what shall we do." Despite this difference in focus, stochastic programming and options theory have a lot in common. Value is often a byproduct of a stochastic program, and options problems incorporate decision making.

The term *real* options theory stands in contrast to *financial* (not imaginary!) options theory and indicates that we are talking about real operational decisions rather than the implicit decisions embedded in a financial security. Like financial options, a large part of the value in operational decisions lies in the possibility to make better decisions in the future.

All option valuations can be formulated as stochastic programs. But there are many problems that lend themselves to stochastic programming where options theory cannot be used.

1.8.1.1 Building an Oil Platform

For a concrete example, assume you are consulting on the construction of an oil platform. One of the issues relates to the size of the platform deck.

The bigger the deck, the larger the platform legs, and the more expensive the platform. By adding extra deck space, you buy an option, however: you make it possible later on to add more equipment. Should the field produce more sand than expected, you could add an extra unit for sand removal. Should the platform produce more oil than expected, you could add a production unit, and maybe you would end up producing so much water that you would need an extra unit for that. So on the one hand you have a certain cost, on the other hand an uncertain gain. You buy the option if the expected gain is higher than the certain cost. The stochastic programming approach to pricing such an option takes a different route than is followed in finance courses.

In stochastic programming, our emphasis will be on the practical side— using the stochastic programming approach makes it much easier to incorporate details of actual operations. On the other hand, in stochastic programming we cannot use the more powerful computational approaches and simplifications made possible by options pricing's use of continuous-time stochastic processes. In real options settings, however, operational details may actually be much more important to solution quality than the representation of the probability space. In financial applications, the use of continuous-time stochastic processes yields quickly obtainable solutions, and that is often highly valued. It must be noted that in many real optional settings, and even some financial ones, the price of ignoring operational details has a great impact on the business value of the solutions.

1.8.1.2 Oil-Well Abandonment

The canonical problem in real options theory is oil-well abandonment. An oil well consumes electricity among other services and produces a profit that depends on the price of oil. As the oil field ages, the quantity of oil declines at a rate that is fairly well understood. Abandoning the oil well costs money and is irreversible for all practical purposes, at least for this problem setting. The question is *when to abandon the well*.

The deterministic approach to this problem is to apply net present value (NPV) analysis. According to NPV analysis, the well should be abandoned when the expected discounted future profits equal the cost of well abandonment. This is correct if there is no uncertainty. However, the price of oil (and of electricity) is variable. NPV neglects the profit that could be obtained if the price of oil shoots up faster than the cost of operations. This potential profit is called the "option value" of the well; for further information on the real options approach, see Sick [51].

For us the concept of option is included in the notion of *recourse actions*, which embody the potential to respond when information has been observed. The additional and very important detail of *when* to execute a recourse action is a topic that lies beyond the scope of this book.

1.8.1.3 Finding Recourse Options

In Sect. 1.4 we discussed robustness and flexibility, and we pointed out how flexibility and robustness related to one another. We showed, for example, how buying a (real) option to face a future uncertainty is a robust action leading to future operational flexibility.

But the options we defined were all very explicit and well defined. These can come about in two different ways: problem knowledge and solution of a stochastic program. Once an option is defined, it can (if feasible) be evaluated using options theory or be entered into a stochastic program as a possible decision. Let us first look at two cases where problem knowledge has been used to define options.

Scheduling and Real Options

Scheduling is an important and economically significant planning problem all on its own. Airlines have large divisions for handling real-time disturbances or shocks in their operations. They cancel flights, reroute passengers, switch crews or planes, deadhead crew, put passengers in hotels, or whatever is necessary to minimize the cost of a disturbance. The cost includes both monetary outlays and bad publicity that impacts future revenues. The costs are substantial, and keeping costs low is a major goal. Substantial effort is invested into developing effective methods to recover from shocks.

Recovery Costs Versus Planned Operating Costs

There is an obvious, but still much overlooked, problem in the foregoing approach. As the ability to find good (deterministic) solutions increases, the planned costs (the costs incurred when everything goes as planned) become lower and lower. However, implicit in the formulations used to find schedules we almost always find the assumption that the world is deterministic. Of course, nobody actually thinks this is the case. But that is beside the point: The models are deterministic, hence leading to solutions with typical knife-edge properties, as discussed on p. 6.

Such solutions normally do not have good expected behavior relative to surprises such as delays and breakdowns. We saw that in our news mix example on p. 4, and it is also outlined in Higle and Wallace [26] and Wallace [55]. Hence, the expected recovery costs might become very high. So the combination of a world-class scheduling department and a world-class recovery department may lead to a mediocre total performance because they fit so badly together.

Real Options in Airline Scheduling

In the airline industry, we now see an increased understanding of these issues. Ehrgott and Ryan [11, 12] have described a model where the robustness variable is extra ground time, and optimization amounts to finding an optimal distribution of ground times. This way the airlines try to maximize the probability that minor delays will have no effects at all. Rosenberger et al. [47] went one step further and analyzed the potential effects of limiting the number of airplane types at certain medium-sized and small airports.

Adding constraints will increase the planned costs. But this makes the schedules, as seen by the customers, much better as they more effectively withstand shocks. The quality comes from the fact that with fewer airplane types, the ability to switch crews and planes on short notice increases, thereby increasing the day-by-day operational flexibility. In addition, this approach reduces maintenance costs. Robustness and flexibility in the airline industry are also discussed in Ball et al. [2].

Telecommunications

From telecommunications we can follow the ideas of robust network designs back to Suurballe and Tarjan [52], who discussed two-connected networks, that is, networks where each pair of nodes is connected by at least two paths not sharing arcs. That makes the network well protected against arc failure since information can be routed along different paths. Even better performance can be obtained by going to three-connected and generally n-connected networks.

What characterizes these ideas from airline scheduling and telecommunications is that they are obtained by problem knowledge. The model defines what we are to check. This is also in line with what options theory offers (although the previously mentioned articles do not relate to options theory), namely, an answer to the question: "Here is an option, what is it worth?" With this ability to find value, we can also compare options. However, options theory cannot be used to *find* options.

Options theory cannot be used to find options, just to evaluate them.

This lack of ability to *find* optional actions is particularly serious. Options theory can never tell the absolute quality of a solution. One may find the best of four options (correctly), but one cannot know if there are substantially better options available. We refer to Chap. 5 for the analysis of a case where defining the *recourse options* is the critical step.

1.8.2 Chance-constrained Models

Chance-constrained models are models that replace a constraint of the type $ax \leq b$ by one that requires the constraint to be satisfied with a certain probability. For example, that demand must be met 95% of the time or that there must be less than 1% chance of going bankrupt. Often these statements really represent what we want. The 1% chance of bankruptcy might be a company policy that we are forced to follow. The same may be true of the 95% chance of being able to deliver—a 95% service level.

We might wish to ask: What is the optimal service level relative to the costs or benefits? Although most businesses find it convenient to measure goals in terms of service level goals, it is important to keep in mind that there must be a between the value of a high service level and the costs of implementation. By using chance constraints we are pushing the process of checking if the present level is indeed optimal into the background.

The problem with a chance constraint is that the optimization will attempt to ignore bad cases unless measures are taken to address them. A service cost minimization model, say, will try to achieve the 95% service level as cheaply as possible. It will therefore try to handle the 95% easiest cases and leave out the 5% most costly ones. The optimization may disregard some investments that could reduce the costs of the remaining 5%. This problem cannot be overcome by parametrically varying the service level requirement. It can only be addressed by bringing all cases into the objective function where the costs and benefits can be compared.

> Solutions to chance-constrained models may have very bad overall performance unless measures are taken to model costs and benefits arising from the cases that are excluded.

There are two major types of chance constraints: individual and joint. In the first case, we put requirements on an individual constraint, like the probability of going bankrupt. In the second case, we make statements about several constraints at the same time. An example could be the requirement that there be a certain probability that the demand for *all* our products is met. We do not discuss algorithms in this book, but we trust you understand that to set up such a model, you must be able to handle dependence in random variables, as discussed in Sect. 1.6.2. The solution needed to obtain a given service level will normally strongly depend on how the individual random variables depend on each other.

1.8.3 Robust Optimization

Robust optimization is an alternative framework for modeling decision problems where some parameters are uncertain. A perusal of the literature will reveal many flavors of robust optimization, so perhaps there is a bit of confusion about what robust optimization means. In this section, we give brief descriptions of the flavors that we know about.

The main use of the term *robust optimization* in recent years has been in application to methodologies in mathematical programming to analyze the dependence of solutions on parameters. A recent volume by Ben-Tal et al. [1] explores this topic in great mathematical detail. Many problems are highly sensitive to particular parameters—including problems that are widely recognized as canonical examples in various fields. General techniques to identify problems with sensitivities and systematically explore the impact of errors in estimating sensitive parameters are a very good *starting point* to address the sorts of issues that we will concern ourselves with in this book.

The term *robust* has been applied to certain model formulations in stochastic programming, in which the *second stage* collects error terms from soft constraints and includes them in the objective, possibly computed using a nonlinear function.

A third flavor, due to Bertsimas and Sim [5], models a simple two-point error distribution for each parameter independently and asks the question: what is the relationship between the objective function value and the probability of finding a feasible solution for this objective value? Let us call this last flavor "stochastic robust optimization" to distinguish it from the other ones.

1.8.3.1 Stochastic robust optimization

The idea is to look at the uncertain variables as enemies that try to make life hard for you. But it is, according to the authors, not very likely that all the enemies will act at the same time. So the authors define a parameter Γ_i, where the indexing is over rows, which controls how many of the uncertain variables in a given row will act against you at the same time. In this way, the optimal solution $\hat{x}$ becomes a function of Γ, such that if you implement $\hat{x}(\Gamma)$, then you are guaranteed that if up to Γ_i uncertain variables in row i work against you at the same time, even if they all take on their most problematic values, your solution is still feasible. This way you can, in principle, trace a frontier between the protection level Γ and the cost corresponding to $\hat{x}(\Gamma)$.

To do this, you obviously must define an interval over which each uncertain variable can vary. The authors use the nominal value plus or minus a constant. Although this is very similar to a support for a random variable, and certainly may be exactly that, in general it is "simply" an interval over which you wish to be protected. The choice of interval will normally be subjective. The authors are not using random variables as such and, in particular, do not use distributions. This is certainly an advantage in the sense that we do not have to specify the distributions. On the other hand, as we will see shortly, this advantage definitely comes at a cost. Also, in our view, if you have partial information about the distribution, it is bad modeling practice not to use it. Even worst-case analysis should relate to that part of the problem you do not understand. But here the modeling philosophies differ. You have to take a stand on this question.

Stochastic robust optimization is both different from and similar to stochastic programming; it has advantages and disadvantages. One advantage that it shares with stochastic programming is that the problem type does not change radically from its deterministic original. Stochastic linear programs are linear programs, just larger. Stochastic robust optimization versions of linear programs are also linear programs, but stochastic robust optimization problems do not grow in size as fast as stochastic programs in the number of stochastic parameters, which is a huge advantage over stochastic programs. On the other hand, the original formulation of stochastic robust optimization problems allows only simple stochastic distributions and does not really address the complex features involved in the choice of distributions such as utility, risk and reward trade-offs, dependence between random variables, and so forth.

Let us look at stochastic robust optimization formulations in the light of robustness and flexibility as defined by us in Sect. 1.4. The goal of these formulations is to create a first-stage decision (these models have only one stage) that is robust in the sense of being able to withstand shocks. However, there are three related problems here.

Achieving Robustness

The first potential problem is that setting up a model with only robust variables (only one stage) may be costly. Very often robustness at one level is connected to flexibility at a lower level to achieve good overall performance. A goods transportation company facing serious randomness in demand may still allow itself a robust (in our terminology) publicized schedule because such a schedule has operational flexibility in the routing of goods. Remember that such a schedule will be of the type "There is an overnight service from New York to Chicago," but it will not say *how* the goods are sent. A schedule that tries to handle all the uncertainty by using robustness, without taking into account typical second-stage variables such as rerouting of goods, will probably be very expensive and based on having far too many trucks available. Planning routes without taking into account the actual operational flexibility is simply too conservative. A reason for the computational simplicity of stochastic robust optimization is exactly the lack of stages.

Since the models do not have stages, all the decisions are robust in our terminology. It is important to be aware of this different use of the term *robust*.

Feasibility

The second problem with many robust formulations is that they lead to solutions that are feasible only with a certain probability, just as for chance constraints. In fact, chance constraints and stochastic robust optimization result in a very similar formulation and interpretation. (Although we do not pursue this thought, it seems that one could view stochastic robust optimization as a nice way of implementing a certain restricted family of chance-constrained problems.) Consequently, many of the observations we made concerning chance-constrained problems apply here, too.

Interval Sensitivity

The third problem is the choice of interval for the uncertain parameters. The solutions you obtain may be very sensitive to these choices, and in most cases they are rather subjective. Normally, supports of random variables in stochastic programs are also subjectively chosen, but the extreme effects are offset by the willingness to use probabilities, expressing that these extreme values are not very likely. Also, when low probabilities are combined with stages, less extreme effects are achieved in stochastic programs than in the robust formulations.

Combining these three issues, we see that we have to be careful with stochastic robust formulations. Not all forms of robustness are economical. The robustness achieved may be very costly in its own right, and the expected performance may be very bad. In repeated operations, it is usually the expected performance we care about.

> As with chance-constrained formulations, stochastic robust formulations may produce solutions with a bad expected performance.

Remember that the purpose of this section was to illustrate how you might want to think about a new approach to find out if the underlying assumptions really fit your needs. Our discussion concerns only one approach. The comments do not generally apply to robust formulations. Whenever you see (or develop) a new approach, you need to be careful and critical about what it really means. Maybe it fits you perfectly, but maybe it does not. That is your responsibility to find out.

1.8.4 Stochastic Dynamic Programming

We have already pointed out that stochastic programming is primarily about transient decision making. Stochastic dynamic programming, on the other hand, is normally focused on long-term steady-state behavior. We define *state variables*, such as inventory levels, present staffing, the weather—whatever is relevant for our problem—and then for each possible *state*, that is, for each possible value of our multidimensional state space, we calculate what should be done. So we end up with *decision rules* or *policies* saying that if the state is so-and-so, we should do so-and-so. This is in contrast to stochastic programs that give us specific suggestions for decisions in specific situations.

> Stochastic dynamic programming formulations focus on long-term steady-state, not transient, behavior.

If you have a problem that fits the dynamic programming framework, then it is a very efficient approach. But dynamic programming is rather limited with respect to what stochastics it can handle, and it cannot handle very complicated constraints and problem definitions.

Chapter 2

Modeling Feasibility and Dynamics

That man is prudent who neither hopes nor fears anything from the uncertain events of the future.
– *Anatole France*

As was illustrated in our News Mix example in Chap. 1, it is not straightforward to pass from a deterministic to a stochastic formulation. We need to rethink the whole model, very often by changing both variables and constraints. Although many reformulations may make sense mathematically, they may in fact be rather peculiar in terms of interpretations. The purpose of this section is to discuss some of these issues, partly in terms of examples. The goal is not to declare some formulations generally superior to others, but rather to help you think carefully about how you rewrite your problems in light of uncertainty.

2.1 The Knapsack Problem

As an example, let us look at the knapsack problem. The problem is simple to write down:

$$\text{maximize} \quad \sum_{i=1}^{n} c_i x_i$$

$$\text{such that} \quad \begin{cases} \sum_{i=1}^{n} w_i x_i \leq b, \\ x_i \in \{0,1\} , \quad i = 1, \dots, n, \end{cases} \tag{2.1}$$

where

c_i is the value of item i
w_i its weight and
b is the capacity of the knapsack

A.J. King and S.W. Wallace, *Modeling with Stochastic Programming*,
Springer Series in ORFE, DOI 10.1007/978-0-387-87817-1_2,
© Springer Science+Business Media New York 2012

The goal is to fill the knapsack with as many valuable items as possible, but without exceeding the weight limit. Of course, w_i might also be viewed as "size," in which case the volume of the knapsack is the capacity in question.

Assume now that the weights are uncertain, so that, in fact, we are facing a vector of random variables $[w_1, \ldots, w_n]$. How are we to interpret this situation? The first question to be asked is always:

What is the inherent stage structure, and how many stages are there?

A clear clue to the stage structure is *when will we learn the weight of an item?* Obvious suggestions are:

1. We learn the weight *of each item just before* we decide whether or not to put it into the knapsack.
2. We learn the weight *of each item just after* putting it into the knapsack.
3. We learn the weight *of the full set of items just after* we decide what items to put into the knapsack.

The first two interpretations can lead to both *inherently two-stage problems* and *inherently multistage problems* with as many stages as there are items. The last interpretation will normally lead to an inherently two-stage formulation, in that we first decide which items to put in and only thereafter observe if they in fact fit. An additional aspect of stage structure is how potential infeasibilities are handled. After all, even though weights are uncertain, the capacity of the knapsack is fixed.

2.1.1 Feasibility in the Inherently Two-Stage Knapsack Problem

Let us list some potential ways of handling these stage-structure questions. For the moment we limit our discussion to the inherently two-stage cases where all items are picked (or listed in a specific order) before we learn anything about their weights.

1. We may require that the chosen set of items must always fit in the knapsack.
2. We may list the items in a certain order and pick them up until we come to one that does not fit. Then we stop. (So the decision is the list.)
3. We may do as described above, but if a later item fits (as it is light enough), then we take it.
4. We may list the items and keep adding items until we have added one that does not fit. We then pay a penalty for the overweight.
5. We may pick a set of items such that if the items do not fit in the knapsack after we have learned their weights, we pay a penalty for the total overweight.
6. We may pick a set of items of maximal value so that the probability that the items will not fit in the knapsack is below a certain level.

There are certainly more variations, but let us stop here for now. You should think about what these cases imply before reading on. One way to structure the analysis is to ask a central question: When do we learn the weights of the items?

In Case 2 above, we list the items and stop putting them into the knapsack when we find one that does not fit. This implies that we learn the weight of an item *before* it is actually put into the knapsack. Case 3 is a variant of Case 2 since it amounts to stopping when an item does not fit and then continuing down the list until we find ones that do fit. Case 4 implies that we need to actually put the item into the knapsack before observing its weight. So the second and fourth cases represent quite different interpretations of when we learn the weights. In Case 5 we learn the weights after we have decided on the selection of items.

Note that putting items into the knapsack is a very passive action in these cases since we have already decided on the order in which items will be picked up (for Case 5 we simply pick them all up). If we want the order to depend on the actual observed sizes, then we end up with an inherently multistage formulation, which we will discuss a bit later.

Cases 2, 3, and 4 result in inherently two-stage models since we define the lists before we start putting items into the knapsack. Stage 1 is to find the list of items, whereas Stage 2 is to passively put them into the knapsack until the stopping rules are satisfied. But to set up the lists, you must anticipate the different situations that can occur. Hence the models will be multistage with an inherently two-stage structure.

Case 5 is also an inherently two-stage formulation, leading to an inherently two-stage model. The model will have only two actual stages as there is no question of ordering the items.

Case 6 results in a chance-constrained formulation that we will discuss shortly.

Case 1, requiring the chosen set of items to always fit in the knapsack, corresponds to a worst-case analysis. Since we need to find a set of items that always fit in the knapsack, we can replace the random variables w_i by their maximal values. Of course, this formulation makes sense only if there is an upper bound on the weights.

The worst-case analysis corresponds to a very "pessimistic" view of the world: we can *never* accept overweight. Whether or not this is reasonable is a *modeling* question. We must look at the situation in which the model is being used and ask ourselves if it is really the case that we cannot handle an overweight item.

If we plan to put an item into the knapsack, is there nothing we can do to get it to "fit"? If the knapsack is a truck, and the items are the loads we plan to send, could we not send some items with the next truck? Maybe we could put a package in the passenger seat? Maybe we could send it by mail?

Requiring feasibility in this way is extremely strong, and we must be sure we really wish to imply a worst-case situation.

Finally, of course, our estimates of the maximal weights may be incorrect. This may lead to an actual situation with overweight, even if the model said it would not happen! Then what will we do? Will the world end? Will the company go broke for sure? Probably not. But if we *can* handle overweight when it really happens, then why did we model the problem as if it could not be allowed to happen under any circumstances? You should really be able to answer these questions if you wish to use worst-case analysis.

2.1.2 Two-Stage Models

So some of the models, while being inherently two-stage, are multistage in nature. Those that are not are the worst-case analysis (which is always particularly risky to use if the worst case is not well defined) and the last two cases—the one with a penalty for total overweight and the one looking at the probability of overweight. Let us first look at a penalty case.

Let S be the set of scenarios describing the uncertainty in an appropriate way. Then we obtain

$$\max \quad \sum_{i=1}^{n} c_i x_i - d \sum_{s \in S} p^s z^s$$

$$\text{such that} \quad \begin{cases} \sum_{i=1}^{n} w_i^s x_i - z^s \le b, & \forall s \in S, \\ z^s \ge 0, & \forall s \in S, \\ x_i \in \{0,1\}, & 1 = 1, \dots, n, \end{cases} \tag{2.2}$$

where d is the unit penalty for overweight. A more general penalty could be a function $f(z^s)$ describing a nonlinear dependence on the total overweight. This model might be good or bad, depending on how well it describes our case. It has, however, a clear interpretation, as it has a clear information structure (we learn about the weights after having decided which items to use), it has a clear description of the goal (to maximize the value of the items selected minus the expected penalty for overweight), and it states what happens if we get it wrong—we pay a penalty. The penalty may mean exactly that, a financial penalty for being wrong. But it may also mean a cost for sending an item with a later truck, the extra cost of using a competitor, or possibly a rejection cost.

This formulation can be viewed as replacing a constraint with a penalty since it can be written as

$$\max_{x_i \in \{0,1\}} \quad \sum_{i=1}^{n} c_i x_i - d \sum_{s \in S} p^s \left[\sum_{i=1}^{n} w_i^s x_i - b \right]_{+}, \tag{2.3}$$

where $[x]_+$ is equal to x if $x \geq 0$, and zero otherwise. We call this a *penalty formulation*.

Case 1 in our listing, the worst-case analysis, can be formulated as ensuring that each item's maximal weight w^{max} will fit into the knapsack:

$$\max_{x} \quad \sum_{i=1}^{n} c_i x_i$$

$$\text{such that} \quad \begin{cases} \sum_{i=1}^{n} w_i^{\mathrm{max}} x_i \leq b, \\ x_i \in \{0,1\}, \qquad 1 = 1, \ldots, n. \end{cases}$$

There is not much to say about this one. It is very pessimistic, and the model is, technically speaking, deterministic. Of course, in some cases, this is exactly what we need, so the model may be appropriate. Note, however, as mentioned above, that this is a very sensitive model unless w^{max} is well understood. Therefore, although mathematically well defined, this model may be pessimistic *and* risky at the same time. So in general, this model is hard to defend. On the one hand, we claim that the items *must* fit in the knapsack; on the other hand, we risk that they do not unless we know w^{max} precisely. The average behavior of the solution coming from this model might be bad, as we do not attempt to control it.

2.1.3 Chance-Constrained Models

Let us pass to Case 6 in our listing, a model that tries to get around the problem of feasibility by requiring that the items fit in the knapsack with a certain probability. The standard model in this case within stochastic programming is a chance-constrained model. It would take the following form:

$$\max_{x_i \in \{0,1\}} \quad \sum_{i=1}^{n} c_i x_i$$

$$\text{such that} \quad \begin{cases} \sum_{s \in W(x)} p^s \geq \alpha, \\ W(x) = \{s : \sum_{i=1}^{n} w_i^s x_i \leq b\}, \end{cases} \tag{2.4}$$

where α is the required probability of feasibility. This model is clear with respect to the objective function and how to treat infeasibilities.

Chance-constrained models say nothing about what happens if we have overweight. In a sense, a chance-constrained problem is a slight relaxation of a worst-case analysis. In the truck example, this model means that we plan which items to put on the truck and we require that our plans work out α

percent of the time. What we do when the items do not fit is not clear. Perhaps we ship them off on another slower channel, and the probability level is in fact a service level for a quick transfer.

If the monetary cost is reasonably well connected to α, we also have controlled the costs. But a danger with this formulation is that the costs associated with lack of feasibility may be connected to the size of the violation, not just the probability, and a chance-constrained model does not have any control over this aspect of the solution.

2.1.4 Stochastic Robust Formulations

Chance-constrained problems (particularly with discrete variables) can be hard to solve. That is especially so for problems more complicated than what we discuss here. Stochastic robust optimization (discussed in Sect. 1.8.3.1) offers an alternative. The worst-case analysis discussed above is a type of robust formulation—we are looking for the most profitable solution that is always feasible. However, there are more sophisticated formulations with a trade-off between loss in income and increase in probability of feasibility. Also in these models we totally disregard what lack of feasibility actually costs. Let us write the model in a way that is reasonably easy to understand, although this is not the format we would use to solve it. Let there be N items available, such that item i has a weight coming from a symmetric distribution over $[w_i - \hat{w}_i, w_i + \hat{w}_i]$, and let Γ be an integer between 1 and N. Let $\mathcal{N} = \{1, 2, \ldots, N\}$. The following formulation will give us the set of items with maximal value under the constraint that if at most Γ of the random weights work against us, we still have a feasible solution, i.e., we can still fit the items in the knapsack, whatever values these weights take. The general probability of feasibility is also high. Bounds are complicated but are given in the underlying paper by Bertsimas and Sim [5]:

$$\max_{x \in \{0,1\}} \quad \sum_{i=1}^{n} c_i x_i$$

$$\text{such that} \quad \begin{cases} \sum_{i=1}^{n} w_i x_i + \psi(x, \Gamma) \leq b, \\ \psi(x, \Gamma) = \max_{S \subset \mathcal{N}, |S| = \Gamma} \sum_{i \in S} \hat{w}_i x_i. \end{cases}$$

As we increase Γ, we get closer and closer to a worst-case situation. Also in this model, there is no statement about what actually happens when items do not fit. As before, that might or might not be a problem.

A major reason these robust models do not control average profits is, of course, that they do not use probabilities at all, just supports of the random variables (or, generally, intervals over which we wish to be protected). In its own right that may be good, but it certainly carries with it potential surprises when solutions are implemented. The combination of considering neither the costs of overweight nor the probabilities thereof is not without potential risks.

Note that the worst-case formulation given earlier and the preceding case with $\Gamma = N$ are not the same, as the x variables are not defined in exactly the same way.

2.1.5 Two Different Multistage Formulations

When making the knapsack problem stochastic, there are two different multistage settings; one is inherently two-stage, the other inherently multistage. In one case, we ask for a decision rule of the following type: items $i_1, i_2, \ldots, i_k$ have been added and the weights turned out to be $w_1, w_2, \ldots, w_k$, so which item should I put in next? This problem is inherently multistage. This is an extremely difficult problem if you require optimality.

The alternative multistage problem is as follows. Give me the (ordered) list of items and follow certain rules for stopping. Here we do not change our minds on the order as we fit them in based on observations, but we take uncertainty into account when setting up the list. This formulation is inherently two-stage. A good test for you is to formulate this latter problem as a decision-tree problem under the assumption that each item has only a limited number of possible weights. Try it!

2.2 Overhaul Project Example

This example is taken from Anderson et al., Sect. 10.4. Let us first repeat the problem and the analysis as given in the reference. We have an overhaul project with five activities, labeled A–E. The example is as given in Table 2.1. The activity-on-arc network for this little project is given in Fig. 2.1.

Table 2.1: Activity names, expected durations, and immediate predecessors

Activity	Description	Immediate predecessor	Expected duration (days)
A	Overhaul machine I	–	7
B	Adjust machine I	A	3
C	Overhaul machine II	–	6
D	Adjust machine II	C	3
E	Test system	B,D	2

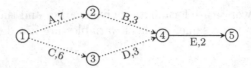

Fig. 2.1: Example network for overhaul example; each arc is labeled with
its activity code and duration

The longest path through this network is given by the sequence of activities
A, B, and E with a completion time of 12. This path is called the *critical
path* as any delay on an activity on this path will delay the whole project.
The partial sequence C–D has a slack of 1, indicating that if either activity
(but not both!) is delayed by 1 day, the project completion will not be delayed.

Suppose that it has become evident that the overhaul project must be
completed within 10 days. With the data presented in Table 2.1, this is not
possible, and the company is willing to invest money to reduce the project
duration. The company may reduce the durations of the activities for a cost.
For each activity, the reduction costs and the maximum possible reductions
are given in Table 2.2.

Table 2.2: Maximum possible reduction and corresponding costs

Activity	Description	Maximal reduction	Cost per day
A	Overhaul machine I	3	100
B	Adjust machine I	1	150
C	Overhaul machine II	2	200
D	Adjust machine II	2	175
E	Test system	1	250

Let y_A denote the number of days by which we reduce the duration of activity A, which will cost $100y_A$. With variables for the other activities similarly
defined, the following linear program can be used to determine the minimum
cost of reducing the project duration to 10 days, as required. The variable x_i
denotes the time at which event i begins. For example, event 4 is the time
when both activities D and B have finished and, hence, E is ready to start.

$$\text{min} \quad 100y_A + 150y_B + 200y_C + 175y_D + 250y_E$$

$$\text{such that} \begin{cases} x_2 \geq x_1 + 7 - y_A, \\ x_3 \geq x_1 + 6 - y_C, \\ x_4 \geq x_2 + 3 - y_B, \\ x_4 \geq x_3 + 3 - y_D, \\ x_5 \geq x_4 + 2 - y_E, \\ x_1 = 0, \\ x_5 \leq 10, \\ y_A \leq 3, \\ y_B \leq 1, \\ y_C \leq 2, \\ y_D \leq 2, \\ y_E \leq 1, \\ x, y \geq 0. \end{cases} \qquad (2.5)$$

The minimal investment necessary to complete the project within 10 days is 350 and is obtained by setting $y_A = 1$ and $y_E = 1$ with all other investments zero. All activities are now on a critical path. This is a typical result from this type of model. Our investment results in a large number of critical paths, and the extreme case is what we observed here: all activities sit on critical paths. The new network is given in Fig. 2.2.

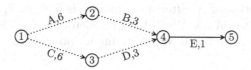

Fig. 2.2: Example network after investments in activity durations; all paths are critical

2.2.1 Analysis

How are we to interpret this result? At a cost of 350 we have reduced the project duration to 10, as required, and all activities have become critical. What does that mean? Does it mean that any delay in any activity will cause a project delay? The answer depends on the nature of the data presented in Tables 2.1 and 2.2.

Suppose that the activity durations given in Figs. 2.1 (before investments) and 2.2 (after investments) are not really deterministic. Although many possibilities exist, let us assume that the given numbers are expected durations and that the distributions are independent and as indicated in Table 2.3. That is, suppose that with the exception of activity E, all activity durations will be their expected values, plus one of three values, $\{-1, 0, 1\}$, with all values being equally likely. Activity E is assumed to have a deterministic duration.

Table 2.3: Probability distribution for activity durations relative to mean

		Probability of deviation from expected value		
Activity	Description	-1	0	$+1$
A	Overhaul machine I	$\frac{1}{3}$	$\frac{1}{3}$	$\frac{1}{3}$
B	Adjust machine I	$\frac{1}{3}$	$\frac{1}{3}$	$\frac{1}{3}$
C	Overhaul machine II	$\frac{1}{3}$	$\frac{1}{3}$	$\frac{1}{3}$
D	Adjust machine II	$\frac{1}{3}$	$\frac{1}{3}$	$\frac{1}{3}$
E	Test system	0	1	0

Given that activity durations are random variables, the project duration is also a random variable. Since this is such an easy network, we can calculate the distribution of the project duration exactly. We begin by calculating the distribution of the duration of activities A and B together and then do the same for C and D. Event 4 will then take place when the last of (A and B) and (C and D) is finished. Finally, we add the duration of E.

Since the duration of A (after investments) is 5, 6, or 7, while that of B is 2, 3, or 4, activities A *and* B will have finished after 7, 8, 9, 10, or 11 days. The distribution can be found by examining all possible combinations of the completion times:

Duration of A	5	5	5	6	6	6	7	7	7
Duration of B	2	3	4	2	3	4	2	3	4
Duration of A *and* B	7	8	9	8	9	10	9	10	11

Each of these events occurs with probability $\frac{1}{3} \times \frac{1}{3} = \frac{1}{9}$. Hence, the distribution for the duration of A *and* B is given by

Duration of A *and* B	7	8	9	10	11
Probability	$\frac{1}{9}$	$\frac{2}{9}$	$\frac{3}{9}$	$\frac{2}{9}$	$\frac{1}{9}$

This is also the duration of C *and* D. Event 4, which corresponds to the start of activity E, occurs when all of the first four activities have finished. Formally, we may state that

Start time for event 4 = max{Duration of A and B, Duration of C and D}

To calculate the distribution of this maximization, we simply look at all 25 combinations of durations of (A and B) and (C and D). The project duration is simply the start time for event 4, plus the duration of activity E (which is 1 day). Thus, we obtain the following distribution for the duration of the project:

Duration of project	8	9	10	11	12
Probability	$\frac{1}{81}$	$\frac{8}{81}$	$\frac{27}{81}$	$\frac{28}{81}$	$\frac{17}{81}$

The expected project duration is therefore 10.642, well above the required duration of 10. In fact, there is a probability of 56 % that the project will take longer than 10. Hence, it seems that our investment of 350 has not brought the duration down to 10. It has not even brought the expected duration down to 10.

At this point you should worry about the sequencing of decisions and about the information that is available as individual decisions are made. Our initial analysis assumed that all activities ended up with their expected duration and that all decisions were made initially. That is, our initial analysis focused exclusively on one scenario that had a probability of $\frac{1}{81}$ of occurring.

2.2.2 A Two-Stage Version

In the preceding discussion, we solved a problem where all activities were assumed to have an average duration in order to find a possible investment. Thereafter, we checked how this solution/investment would behave in a random environment. But when searching for the solution, we did not consider the uncertainty in the activity durations. As a result, one of the effects was that we did not even achieve an expected project duration of 10.

Let us now reconsider the investment, this time recognizing the uncertainty in the activity durations. Suppose that the investment must be determined before the actual durations can be known. This is inherently two-stage. Let d_A^s be the duration of activity A in scenario s, and let the other durations be defined accordingly. Since the duration of activity E is not subject to uncertainty, we do not define such an entity for this activity. Let the investment variables, y, be defined as before, and let x_i^s be the time of event i if scenario s occurs. We have now $3^4 = 81$ equally likely scenarios corresponding to the 81 possible realizations of durations of activities A–D.

A question that occurs at this time is what we are to mean by the project taking 10 days. Is it going to be an average of 10 days, or is 10 days a hard constraint? Or maybe 10 days is the goal, but, at a penalty, we are allowed to be late? In reality, of course, this type of constraint is not hard. We can never guarantee that a project cannot be late. We could certainly find an investment that, with our 81 scenarios, guaranteed that we were never late, but reality will always be different. Hence, let us instead assume that if we are late, a penalty of 275 per day is incurred.

Note first that if we had added the possibility of being late at a penalty of 275 to the deterministic problem, the solution would not have changed, as it is cheaper to invest in reducing activity durations than to pay the penalties. Also, note that the expected cost associated with the deterministic solution is now the initial investment of 350 plus an expected penalty for being late of 210, at a total of 560. So the initial cost estimate of 350 was far off the actual cost.

The following two-stage model will minimize the expected cost of achieving a project duration of 10, *provided all investment decisions are made before the project starts and we are allowed to be late.* Lateness in scenario s is measured by t^s.

$$\min \quad 100y_A + 150y_B + 200y_C + 175y_D + 250y_E + \frac{275}{81}\sum_{s=1}^{81} t^s$$

such that
$$
\begin{cases}
x_2^s \geq x_1^s + d_A^s - y_A & \text{for all } s, \\
x_3^s \geq x_1^s + d_C^s - y_C & \text{for all } s, \\
x_4^s \geq x_2^s + d_B^s - y_B & \text{for all } s, \\
x_4^s \geq x_3^s + d_D^s - y_D & \text{for all } s, \\
x_5^s \geq x_4^s + 2 - y_E & \text{for all } s, \\
x_1^s = 0 & \text{for all } s \\
x_5^s - t^s \leq 10 & \text{for all } s, \\
y_A \leq 3, \\
y_B \leq 1, \\
y_C \leq 2, \\
y_D \leq 2, \\
y_E \leq 1, \\
x, y \geq 0, \\
t^s \geq 0 & \text{for all } s.
\end{cases}
\tag{2.6}
$$

What distinguishes Problem (2.6) from Problem (2.5) is found in the representation of the constraint on the project duration. In (2.5), this was represented by

$$x_5 \leq 10.$$

In (2.6), we have

$$x_5^s - t^s \leq 10,$$

indicating that the duration should be at most 10 but can be allowed to be longer. Furthermore, all constraints that are used to calculate the event times, such as

$$x_2^s \geq x_1^s + 7 - y_A,$$

are now indexed by scenario. Solving this we find $y_A = 1$ as the only investment. The cost is 100. The expected penalty for delays is as high as 455, yielding a total of 555, a reduction of 5 from the expected value of the deterministic solution. (As this is just an artificial example, the numbers themselves are not important; the main point is that we obtain a different and cheaper solution.)

What happened here is that as we realized explicitly that the world was stochastic and that delays were in fact feasible (at a cost), we ended up investing much less initially to reduce the project duration. Instead, we preferred to pay the penalty for being late. With $y_A = 1$ as the only investment, there is a probability that event 4 will take place later than time 8 (so that we finish the whole project later than 10) of 89 %. Hence, the penalty is incurred in 89 % of the cases. As 89 % of 275 equals less than 250 (the unit investment cost in activity E), we prefer the penalty cost. So in this case, there was flexibility in *waiting*. Instead of securing the project duration initially, it was better to wait and see what happened.

This formulation is an inherently two-stage formulation, leading to a two-stage model, as we first make investments, then observe the activity durations, and finally (Stage 2) calculate the project duration and delay costs.

2.2.3 A Different Inherently Two-Stage Formulation

The ideal formulation of this project scheduling problem is to take into account the actual float of information. Initially, we must decide on y_A and y_C. Then, y_B is decided when activity A is finished and y_D when activity C is finished. However, we do not know which of these events will occur first. Modelingwise, this creates a lot of difficulty if we are to formulate the problem as a stochastic programming problem. It makes us unable to define stages. Stage 1 is to define y_A and y_C, but what is Stage 2? It is to define y_B if A finishes before C, but to define y_D if B finishes before A. And which of these will happen first depends on both the randomness and our first-stage decisions. Stage 4 is in any case to determine y_E.

Hence, let us analyze a somewhat simpler case. Let us assume that investments in activities A–D must be determined initially, but that activity E can wait until the activity is to start. This will make y_E scenario dependent. In addition, we can be late at a penalty.

$$\min \quad 100y_A + 150y_B + 200y_C + 175y_D + \frac{250}{81}\sum_{s=1}^{81} y_E^s + \frac{275}{81}\sum_{s=1}^{81} t^s$$

such that
$$\begin{cases}
x_2^s \geq x_1^s + d_A^s - y_A & \text{for all } s, \\
x_3^s \geq x_1^s + d_C^s - y_C & \text{for all } s, \\
x_4^s \geq x_2^s + d_B^s - y_B & \text{for all } s, \\
x_4^s \geq x_3^s + d_D^s - y_D & \text{for all } s, \\
x_5^s \geq x_4^s + 2 - y_E^s & \text{for all } s, \\
x_1^s \quad = 0 & \text{for all } s, \\
x_5^s - t^s \leq 10 & \text{for all } s, \\
y_A \quad \leq 3, \\
y_B \quad \leq 1, \\
y_C \quad \leq 2, \\
y_D \quad \leq 2, \\
y_E^s \quad \leq 1 & \text{for all } s, \\
x, y \quad \geq 0, \\
t^s \quad \geq 0 & \text{for all } s.
\end{cases}$$

(2.7)

Based on what we have learned so far, it is a challenge to guess what the solution will be. Without calculations, you should be able to see that the investment will be $y_A = 1$ as the only investment. Furthermore, since investments in E can be delayed until we are ready to start the activity, we will choose to invest in E if we start later than time 8. This is so since the cost of investing in E is lower than the penalty cost of being late. If we are ready to start activity E later than time 9, we will invest in a one-unit decrease in E and take the rest of the delay as a penalty. The total expected cost is down to 533.

2.2.4 Worst-Case Analysis

An obvious way to make sure the duration is at most 10 (in fact, the only way) is to perform a worst-case analysis. We then resolve (2.5), but with maximal durations rather than average durations. Hence, the durations of activities

A, B, C, and D will increase by one. The result will be $y_A = 3$, $y_B = 0$, $y_C = 0$, $y_D = 2$, and $y_E = 1$, with a total cost of 900. Of course, in this case the expected delay cost is zero. But be aware that this is assured only if our description of uncertainty is correct. Hence, worst-case analysis can be both conservative and risky at the same time; we are careful, pay a lot to be *sure* that we are always feasible, but then, due to errors in estimating the data, we are not so sure after all. Worst-case solutions do not handle measurement errors.

2.2.5 A Comparison

There are a few points to be made here. The first concerns feasibility. In the deterministic model, it was reasonable and meaningful to say that we had to finish in ten time periods. But if we kept that requirement in the stochastic setting, we were brought to a worst-case analysis. If we have an inherently two-stage model (i.e., all investments are made before the project is started), the only way to guarantee a duration of ten time periods is to plan as if everything were against you. We saw that the cost would be 900. Very often, such strict interpretations of feasibility are not reasonable. Instead, it is necessary to ask if a constraint is *really* hard? Very often the answer is no. If the softness that is therefore brought into the model can be described by penalties, then we end up with a *recourse* model. We gave two examples of such models. In a two-stage setting with penalties, we ended up with expected costs down from 900 to 555. If, in addition, we allowed the second stage to also contain a genuine investment, the expected cost dropped to 533. The latter drop is simply caused by the new investment opportunity's being cheaper than the delay cost. If we solved the deterministic model, it claimed that the investment cost would be 350, whereas, in fact, the total expected cost was 560. With a strict interpretation of feasibility, the deterministic solution was infeasible with a probability of 0.56.

In addition to the issue of feasibility, we observe that there is again a value for delaying decisions. We saw that the total expected cost depended on how we defined these possible delays. We cannot say that one model is better than the other unless we actually know the real decision context. But we see how the modeling choices affect decisions and costs.

2.2.6 Dependent Random Variables

In Sect. 2.2.1, we assumed that all the random durations were independent. We then found that using the investment from the deterministic model $y_A = y_E = 1$, the expected cost was not 350, as indicated by the deterministic model, but rather 560 if the late penalty were set at 275 per day. The probability of being late was as high as 56 %. If the durations had instead been correlated, the deterministic model would have been the same, hence the investments would have been the same, but the expected costs would have been different, and so would the probability of delays.

Let us see what would happen if activities A and B were perfectly negatively correlated and activities C and D perfectly positively correlated. These are, of course, extreme assumptions, but they serve to illustrate a few points.

First, the duration of A and B would be deterministically equal to 9. The perfect negative correlation (correlation coefficient of -1) has removed the uncertainty on that path. As always, a negative correlation has helped us control the variation. A negative correlation of -1 is, of course, rather special (as it implies we have only one random variable, not two), but any negative correlation between the durations of A and B would reduce overall uncertainty and be useful to us.

> Negative correlations are always potentially useful, and you should think carefully about how they might help you.

For the other path, a perfect positive correlation would cause the duration of C and D to be 7, 9, or 11, each with a probability of $\frac{1}{3}$. This is worse than in the deterministic case in the sense that while the probability of being above 9 (causing the project to have a duration above 10) has stayed at $\frac{1}{3}$, the expected delay, given that there is a delay, has increased. Here we see that a positive correlation has caused us trouble, as it normally does.

So the starting time for event 4 is again the maximum of the duration of these two paths plus one, leading to the following distribution for the project duration:

Project duration	10	12
Probability	$\frac{2}{3}$	$\frac{1}{3}$

So the probability of being late is now 33 % and the expected project duration 10.67. The expected cost is now 533, down from 560.

The point of this is to understand that there is not just the question of stochastic/deterministic but also *which* stochastic model we are facing. Correlations (and other measures of covariation) have no counterparts in deterministic modeling, making several rather different stochastic settings being represented by the same deterministic model.

So the question is not that one type of stochastics (like the uncorrelated durations we started out with) is "better" than others. Rather, it is that the effects of solving a deterministic model depends on the stochastic setting. We saw that when we changed from uncorrelated random variables to correlated ones (in this one specific way), the expected costs went down, the expected duration went up, and the probability of delay did not change. Again, this is not good or bad; it is simply an observation telling us that issues that do not even come up in deterministic modeling, such as covariation, can be important in valuing the actual performance of a deterministic model.

2.2.7 Using Sensitivity Analysis Correctly

In this example, we assumed that there was a penalty cost of 275 per day if we were late. We simply used it as a number. Is that appropriate in the light of the discussions in this book? You may want to think about that question for a few minutes!

If the number is a specific estimate of the cost of delay based on market activities, like a contractual penalty, *and nothing else*, such as lost image, we may face two situations:

- If this is an estimate of what the future value of the penalty will be, based on its present value, then it should be treated as a random variable, and our approach is not valid in light of our own discussions: we have used expected values instead of the actual distribution.
- If this is a known entity, which we know will not change during the life of the project, then our approach is indeed valid.

But most likely, even if there is a contractual penalty for lateness, there will also be a question of lost goodwill, lost reputation. And the size of that loss is not really known; it is anybody's guess. So 275 is our guess, our chosen value for the overall costs. But it is a guess, a choice, not because we are facing a random variable but because it is up to us to define the penalty: the penalty is in its own right a policy parameter for the company. If that is the case, our approach is appropriate, but it should possibly be accompanied by a parametric analysis on the level of the penalty.

So the right answer depends on the setting. However, most likely, this is a case where parametric analysis is appropriate because the penalty is like a decision variable: it is up to us to set it. We leave it to you to check what would happen if the penalty were slightly different from 275 (in particular a bit higher).

2.3 An Inventory Problem

Let us turn to another very classical model. We are responsible for a production and inventory system where for the next T periods we know the demand. For practical reasons, we do not like production to vary too much from one period to the next, so we have defined an upper bound on changes in production levels ℓ. We are in a setting where demand must be met, but it can be met from outside sources (which means, technically speaking, that we allow demand to be rejected *inside* the model). We have formulated the following inherently multistage production and inventory model for our situation, not taking uncertainty in demand into account.

$$\min \quad \sum_{t=1}^{T} (c_t x_t + f_t I_t + b_t u_t)$$

$$\text{such that} \quad \begin{cases} I_t - I_{t-1} = x_t - d_t + u_t, & t = 1, \ldots, T, \\ |x_t - x_{t-1}| \leq \ell, & t = 2, \ldots, T, \\ x_t, u_t, I_t \geq 0 & t = 1, \ldots, T, \end{cases}$$

with I_0 given. Here I_t is the inventory at the end of period t, x_t the production of period t (determined at the start of the period), c_t the production costs, f_t the unit inventory costs, and d_t the demand in period t. The variable u_t represents external orders or the impact of lost sales measured by b_t. Of course, there are other versions of this model, representing backorders, for example. But the basic model that requires production plus inventory to satisfy demand is common. The first inequality expresses our requirement that production must not change by more than ℓ from one period to the next.

Note that such an inequality can be written linearly as

$$x_t - x_{t-1} + w_t - z_t = 0,$$
$$0 \leq w_t, z_t \leq \ell.$$

However, we will continue to use the absolute value formulation for easier reading.

2.3.1 Information Structure

The first question is always what is random and when do we learn the value of the random variables? Production costs, inventory costs, demand, and production volumes can all be random. For simplicity in this discussion we assume that the only relevant random variables are the demands. From the perspective of stages in stochastic programs, and the corresponding modeling, we have two major choices with respect to when demand becomes known:

1. Demand for a period becomes known before production for that period is determined.
2. Demand for a period becomes known after production for that period is determined.

Neither of these is better than the other, and probably both are incorrect since most likely we learn little by little. But we need to make a choice modelingwise, so let us assume we learn the demand before production is determined. The time line of our interpretation can be found in Table 2.4.

Table 2.4: Time line for our interpretation of the inventory model

	$t=1$		$t=2$		$t=3$	$\cdots$	$t=T-1$		$t=T$	
x_0, d_1	x_1, u_1	d_2	x_2, u_2	d_3	x_3, u_3	$\cdots$	x_{T-1}, u_{T-1}	d_T	x_T, u_T	
I_0			I_1		I_2	$\cdots$		I_{T-1}		I_T

Let $\mathcal{S}$ be a scenario tree describing how d develops randomly over time. (Chap. 4 discusses the making of such trees.) A path of demands in $\mathcal{S}$ is called a scenario, is expressed by $d^s = (d_1^s, \ldots, d_T^s)$, and occurs with probability p^s. Similarly, production and inventory are described by x_t^s and I_t^s. We require that all variables be *implementable* (also called nonanticipative—if you can pronounce that), in the sense that when two scenarios s and σ are such that $d_t^s = d_t^\sigma$ for $t = 1, 2, \ldots, \tau$, then $x_\tau^s = x_\tau^\sigma$ and $I_\tau^s = I_\tau^\sigma$. The reason is that since the two scenarios are indistinguishable when the production for period τ is made, the production and inventory decisions must be the same.

It is worth a few minutes of your time to be sure you understand this way of formulating a stochastic program. Earlier, in Fig. 1.1, you saw a (simple) scenario tree. That tree branched each time we learned something, and each node in the tree had its own decision variables. A scenario is a path in that tree, and the scenarios interact directly by all scenarios sharing the top node (Stage 0). But there is an alternative formulation. It is sometimes chosen because it provides a better way to outline the problem structure, sometimes because we plan to use a solution method that is based on the formulation. For this book, we want to emphasize that it might at times be easier to read for people not used to stochastic programs. Consider Fig. 2.3, where we challenge you to fill in what is missing.

As you can see, each scenario, represented by a column in Fig. 2.3, has its own set of variables. The first line shows information that just became known when we reached this node (incoming inventory and this period's demand), the second row what must be determined (production and external orders). That means that the basic part of the model is as in the deterministic situation—all information is known. But in addition we impose some requirements. These are what we call implementability or nonanticipativity constraints. Consider the first row in the figure. All boxes have been connected by horizontal lines. That means that decisions in those eight boxes must take on the same values. In particular:

$$x_1^1 = x_1^2 = \cdots = x_1^7 = x_1^8$$

and

$$u_1^1 = u_1^2 = \cdots = u_1^7 = u_1^8.$$

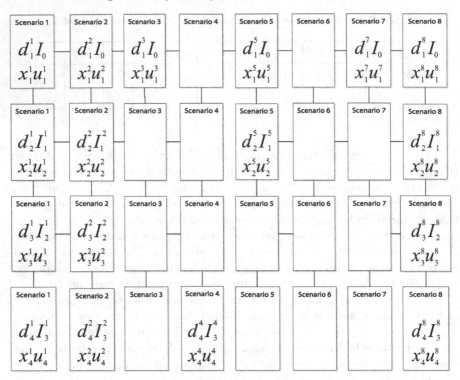

Fig. 2.3: Information structure of the inventory problem outlined scenario by scenario for $T = 4$. A column represents a scenario. Two boxes with a horizontal line between them must have the same values for all the variables

The meaning of this is that since at this point in time we cannot know which of the eight scenarios we are on, we cannot allow decisions to depend on scenarios. Remember that we assumed that demand became known *before* production decisions were made. Hence, in fact, we have assumed that

$$d_1^1 = d_1^2 = \cdots = d_1^7 = d_1^8$$

and, more obviously, that incoming inventory I_0 is the same for all scenarios.

Here you can see the meaning of the two, possibly slightly peculiar, terms *implementability* and *nonanticipativity*. A decision is implementable, i.e., can be implemented, if it does not depend on values that are not yet revealed. "Buy IBM stock now if its value goes up next month" is *not* implementable, while "Buy IBM stock now" is. In the same way, "Buy IBM stock now if its value goes up next month" is anticipative, i.e., it anticipates (uses) information that is not yet known. Nonanticipativity then means that a decision does not use such unavailable information.

The term nonanticipativity is most used. But be a bit careful about how you understand it. The whole point of stochastic programming is to be anticipative, that is, to look into the future and consider what *might* happen. However, you must not anticipate what *will* happen.

In the second row in the figure, the four left nodes are connected, as are the right four. This implies that since $d_2^1 = d_2^2 = d_2^3 = d_2^4$, all the first four scenarios must have the same second-stage decisions since we cannot know which of them we are on. The same applies to the right four nodes. Then, in the third row, only two and two nodes are connected, and then at the end, you know exactly which scenario you are on.

So to say that the variables are implementable, or nonanticipative, is to say that they must satisfy the information structure inherent in Fig. 2.3. A straightforward formulation then becomes (with $x_0^s = x_0$ given for all $s \in \mathcal{S}$):

$$\min \quad \sum_{s\in\mathcal{S}} p^s \sum_{t=1}^{T} (c_t x_t^s + f_t I_t^s + b_t u_t^s)$$

such that
$$\begin{cases}
I_t^s - I_{t-1}^s = x_t^s - d_t^s + u_t^s, & t = 1,\ldots,T; s \in \mathcal{S}, \\
|x_t^s - x_{t-1}^s| \le \ell, & t = 2,\ldots,T; s \in \mathcal{S}, \\
x_t^s, I_t^s, u_t^s \ge 0, & t = 1,\ldots,T; s \in \mathcal{S}, \\
x_t^s, I_t^s, u_t^s \text{ implementable} & t = 1,\ldots,T; s \in \mathcal{S}.
\end{cases} \quad (2.8)$$

Hopefully you can see that, although Fig. 2.3 is perhaps a bit involved until you get used to it, the advantage of being able to write down the problem in terms of scenarios can be substantial in terms of clarity. This is especially true for multistage problems.

2.3.2 Analysis

Let us now see what this model implies, particularly relative to the deterministic version. First, the inventory constraint, combined with nonnegativity on inventory, has turned the model into something like a worst-case model: however high the demand, we *must* be able to satisfy it. As a modeler, you might ask: was this what we wanted? Of course, the deterministic model also had this property, but since demand (most likely) was set at its mean value, the actual effect was less dramatic.

On the other hand, in light of this we might ask what the deterministic model means (unless the modeler really thinks the world is deterministic). The model requires that we *must* meet average demand and does not allow for *any* violation of that requirement. But at the same time, obviously, if the results from the model are to be used in reality, then a shortage will occur. The model does not say anything about what will happen then.

Observe also that while demands up to the mean must be met at any cost, demands above that level do not matter at all! Such an arbitrary setup can lead to rather strange results in terms of behavior in the real world. So we see a deterministic model that makes good sense in its own setting—a deterministic world—but that becomes rather strange when we start to think about its use in a real setting. The problem is that deterministic models often do not answer major questions about what to do under difficult conditions, simply because they assume them away.

We also should look at the constraint on variation in production. As demand now varies more than in the deterministic model, it is now a much more serious constraint. The model might decide on large inventories simply to facilitate cases (possibly with very low probabilities) where demand changes a lot from one period to the next. Was that really the understanding of the underlying problem? Especially the total combination of all constraints results in a very special situation: limited ability to change production plus an absolute need to deliver will give very large inventories, most likely at a level that users will consider unreasonable.

Of course, the foregoing situation might be the right one. It is not our point to say it cannot be. But most likely the starting point will be a situation where shortages (whether they result in lost sales or backorders) are considered bad and where large changes in production should be avoided. But if demand suddenly increases dramatically, it will likely be possible to increase production by more than ℓ. And if inventory is empty, most likely the company does not go directly out of business. There are solutions.

Ways around this could be to put the constraints of smoothness in production into an objective function, with a penalty for all changes or all changes above ℓ. Also, we should think seriously about modeling backorders or lost sales unless we *really, really* must satisfy all possible demand. But if we do not, we face another problem: are we sure our estimates of maximal demands are correct? Most likely we are not. And if so, the model is a bit shaky: while we require feasibility at any cost, at the same time, in reality, shortages might occur. This does not represent good modeling.

2.3.3 Chance-Constrained Formulation

We can also imagine probabilistic constraints here. With discrete distributions, one (but certainly not the only) possibility is to replace $I_t \geq 0$ by

$$\sum_{s \in \mathcal{W}_t(x)} p^s \leq 1 - \alpha, \quad \mathcal{W}_t(x) = \{s | I_t^s < 0\},$$

expressing that at most $1-\alpha$ parts of the time inventory may be negative, time period by time period, in the tree. Apart from obvious problems of solving such a model (remember that I_t depends on all previous demands and production decisions), this model does not really soften the original hard constraints of zero inventory levels. It simply moves them to certain negative levels and then requires that these new limited negative levels must be achieved at any cost, whereas beyond that level, costs are of no concern at all. This is a rather general observation. Chance constraints may appear to be softer than the original constraints. But in general they are not; they are simply different, usually looser, but not softer.

In this example we expressed the chance constraints period by period, so the requirement is not history dependent. An alternative formulation would express the inventory constraint node by node in the scenario tree, rather than period by period.

2.3.4 Horizon Effects

Many problems, particularly those that are inherently multistage, in fact have infinitely many stages. By that we do not imply that the problems cover infinitely long time intervals but that it is not known when the problems end.

A production company knows, of course, that eventually a certain product will go out of production, but they do not know when, so they plan as if the production will go on forever. A financial investor will, sooner or later, go out of business, but he plans as if he will not. That is the nature of most decision problems. We really have no choice but to treat these problems as if they had infinitely many stages. But there is no way we can, technically speaking, handle models with infinitely many stages. We need to make some kind of simplification or approximation to make the model have finitely many stages—if possible, only two.

There is also a modeling-related reason for this: As the number of stages gets very large, we approach a kind of steady-state situation. But as was pointed out previously, stochastic programming is about transient behavior, not steady-state behavior. We are interested in what to do *now*, not what to do when (if) we eventually reach a steady state. So at best we need to represent the steady state in the model to obtain the right transient behavior. But what actually to *do* once we get there is not at all our focus.

2.3.5 Discounting

In problems with many time stages stretching into the distant future, we want to account for the fact that a payment in the future is less valuable than a payment today. For example, we could take a dollar today and invest it in a secure deposit account that pays interest, say, r during each period t. One dollar invested today in this account for t periods would hold

$$\prod_{\tau=1}^{t} (1 + r).$$

The inverse of this value represents the amount we would put into the deposit account today in order to receive exactly one dollar at time t. This ratio $\delta = (1 + r)^{-1}$ is called a *discount rate*. As time passes, discounting progressively lowers the value of a future dollar. For instance, at $15\,\%$ interest rate today's value of a dollar to be received 5 years from now is $\$0.50$, and 10 years from now it is $\$0.25$.

The *discounted present value* of a future cash stream $f = [f_1, \ldots, f_T]$ is

$$\sum_{t=1}^{T} \delta^t f_t.$$

This formula simply adds up the various amounts we would have to put into our deposit account to generate each term of the payment stream f. As time goes on, the discounting reduces the impact of the future on present decision making and imposes a kind of "significance" horizon on comparisons between future cash streams. Discounting is a very common practice in planning problems, but we do not intend for you to always accept such smoothing when it appears. For example, is it sensible to discount global warming or nuclear conflict? It does not take much of a discount factor to smooth away the impact of serious events that are far away in time.

2.3.6 Dual Equilibrium: Technical Discussion

So now suppose in the production-inventory problem that there is a time point t in the future when we will not be too concerned about randomness. We could just as well use mean values for the costs and demands. Even if we do not necessarily know what these will be, we do not want the model to assume that they are *zero* since this could introduce strange incentives into the preceding periods. In stochastic programming, for numerical reasons, we cannot use lots and lots of time stages to gently smooth out the impact of a shock like that, but we can develop an approach that approximates this kind of smoothing.

Let us look at the optimization problem to be solved at this final horizon stage t. The data we have are the decisions from the past stage $t - 1$ and our best guesses as to the steady-state costs and demands. Let us also introduce a time discount factor $0 < \delta < 1$ that applies to the costs in the horizon objective function:

$$c_{t+\tau} = \delta^\tau c_t,$$
$$f_{t+\tau} = \delta^\tau f_t, \tag{2.9}$$
$$b_{t+\tau} = \delta^\tau b_t.$$

This discounting will have the effect of smoothing the impact of errors in horizon estimates, which is what we desire for this model:

$$\min \sum_{\tau=0}^{T-t} \delta^\tau \left(c_{t+\tau} x_{t+\tau} + f_{t+\tau} I_{t+\tau} + b_{t+\tau} u_{t+\tau} \right)$$

such that
$$
\begin{cases}
I_{t+\tau} - I_{t+\tau-1} = x_{t+\tau} - d_{t+\tau} + u_{t+\tau}, & \tau = 0, \ldots, T-t, \\
x_{t+\tau} - x_{t+\tau-1} \le \ell, & \tau = 0, \ldots, T-t, \\
x_{t+\tau} - x_{t+\tau-1} \ge -\ell, & \tau = 0, \ldots, T-t, \\
x_{t+\tau}, I_{t+\tau}, u_{t+\tau} \ge 0 & \tau = 0, \ldots, T-t.
\end{cases}
$$

$$(2.10)$$

Dual equilibrium (due to Grinold [18]) simplifies the horizon problem by equalizing all future undiscounted shadow prices. Of course, this means that the shadow prices are not going to be set at the level that guarantees constraint satisfaction, so in effect we are choosing to overlook feasibility violations in order to simplify the horizon problem. This may not be the correct approach in every modeling situation, so let us think about what it means in the production-inventory problem. The constraints model the satisfaction of demand while restricting production variations from one period to another. If the constraints are not satisfied in some period past the horizon, should we be concerned? This is problem dependent, of course, but let us proceed as if this were a reasonable assumption.

To implement the dual-equilibrium assumption in (2.10), we start by denoting the dual multipliers (shadow prices) for the first, second, and third constraints in each horizon stage by y_τ, z_τ^+, and z_τ^-, respectively. The first step in the dual-equilibrium approach assumes that the dual multipliers all take the form

$$
\begin{aligned}
y_\tau &= \delta^\tau y_t, \\
z_\tau^+ &= \delta^\tau z_t^+, \\
z_\tau^- &= \delta^\tau z_t^-,
\end{aligned}
$$

$$(2.11)$$

which, as we can see, has the effect of equalizing the undiscounted shadow prices for each time stage. If these dual multipliers were actually optimal, then the primal solution for (2.10) could be recovered by optimizing the following objective function:

$$
\begin{aligned}
\sum_{\tau=0}^{T-t} \delta^\tau \Big[&\left(c_t x_{t+\tau} + f_\tau I_{t+\tau} + b_t u_{t+\tau} \right) \\
&+ y_t \left(I_{t+\tau} - I_{t+\tau-1} - x_{t+\tau} + d_{t+\tau} + u_{t+\tau} \right) \\
&+ z_t^+ \left(x_{t+\tau} - x_{t+\tau-1} - \ell \right) \\
&+ z_t^- \left(x_{t+\tau} - x_{t+\tau-1} + \ell \right) \Big].
\end{aligned}
$$

$$(2.12)$$

Of course, these dual-equilibrium multipliers are not likely to be the optimal ones for (2.10). For one thing, there are only three dual multipliers left, so the model only has three constraints! The next step in the dual-equilibrium procedure is to collect terms in expression (2.12) to see what these constraints are.

To see the constraints, we need to add up all three sets of terms that involve the three dual multipliers (y_t, z_t^+, z_t^-) multiplied by the discount factor δ^τ. To simplify these expressions, let us introduce "integrated" primal variables and right-hand sides as follows:

$$d_t^* := \sum_{\tau=0}^{T-t} \delta^\tau d_{t+\tau},$$

$$\ell_t^* := \sum_{\tau=0}^{T-t} \delta^\tau \ell,$$

$$x_t^* := \sum_{\tau=0}^{T-t} \delta^\tau x_{t+\tau},$$

$$I_t^* := \sum_{\tau=0}^{T-t} \delta^\tau I_{t+\tau},$$

$$u_t^* := \sum_{\tau=0}^{T-t} \delta^\tau u_{t+\tau}.$$

What can these integrated expressions mean? Interpreting these expressions really gets us to the heart of the dual-equilibrium approach. By assuming the form (2.11) for the dual-variables structure, we are essentially going to add up (or integrate) all the future demands and production constraints out to the horizon and then choose production, inventory, and external order decisions that satisfy the integrated demand and production constraints. The "integration" incorporates the discount factor, as it should, in order to appropriately scale the impact of the future periods back to the horizon period.

Using the integrated primal variables and right-hand sides you can verify that (2.12) transforms into the following expression:

$$
\begin{aligned}
c_t x_t^* &+ f_t I_t^* + b_t u_t^* \\
&+ y_t \left(I_t^* - \delta(I_{t-1} + I_t^*) + \delta^{T-t} I_T - x_t^* + d_t^* + u_t^* \right) \\
&+ z_t^+ \left(x_t^* - \delta(x_{t-1} + x_t^*) + \delta^{T-t} x_T - \ell_t^* \right) \\
&+ z_t^- \left(x_t^* - \delta(x_{t-1} + x_t^*) + \delta^{T-t} x_T + \ell_t^* \right).
\end{aligned}
\tag{2.13}
$$

We performed the following calculation:

$$\sum_{\tau=0}^{T-t} \delta^\tau x_{\tau-1} = \delta(x_{t-1} + x_t^*) - \delta^{T-t} x_T.$$

Is that really correct? We think it is, but it will help you to understand what is going on if you try to verify it yourself. If we assume that T is so large that

δ^{T-t} is practically zero and drop the asterisks on the primal variables, then the dual-equilibrium horizon problem is

$$\min \quad c_t x_t + f_t I_t + b_t u_t$$

$$\text{such that} \quad \begin{cases} (1-\delta)I_t - \delta I_{t-1} = x_t - d_t^* + u_t, \\ |(1-\delta)x_t - \delta x_{t-1}| \le \ell_t^*, \\ x_t, I_t, u_t \ge 0. \end{cases} \qquad (2.14)$$

The variables (x_t, I_t, u_t) in the horizon problem can be interpreted as the amounts required to satisfy the integrated demand and production constraints out to the horizon.

Dual equilibrium is a general approach to the modeling of horizon constraints. It can be applied to many types of problems. However, like any general approach, there are assumptions made and the solution must be checked to understand if the assumptions are reasonable.

So, in total, we end up with the following extension of (2.8). Notice that T has changed interpretation from the development of the dual-equilibrium horizon model. T is now the period in which we apply the horizon model.

$$\min \quad \sum_{s \in \mathcal{S}} p^s \sum_{t=1}^{T} (c_t x_t^s + f_t I_t^s + b_t u_t^s)$$

$$\text{such that} \quad \begin{cases} I_t^s - I_{t-1}^s = x_t^s - d_t^s + u_t^s, & t = 1, \ldots, T-1; s \in \mathcal{S}, \\ (1-\delta)I_T^s - \delta I_{T-1}^s = x_T^s - d_T^* + u_T^s, & s \in \mathcal{S}, \\ |x_t^s - x_{t-1}^s| \le \ell, & t = 2, \ldots, T-1; s \in \mathcal{S}, \\ |(1-\delta)x_T^s - \delta x_{T-1}^s| \le \ell_T^*, & s \in \mathcal{S}, \\ x_t^s, I_t^s, u_t^s \ge 0, & t = 1, \ldots, T; s \in \mathcal{S}, \\ x_t^s, I_t^s, u_t^s \text{ implementable}, & t = 1, \ldots, T; s \in \mathcal{S}. \end{cases}$$

As before, "implementable" refers to the structure discussed in Fig. 2.3.

2.4 Summing Up Feasibility

The three examples, together with the news vendor example from Sect. 1.2, show many different aspects of modeling stochastic decision problems. However, much of the discussion, one way or another, concerns feasibility. Let us try to sum up what we have seen.

What we notice here is that although many deterministic models can be made stochastic, the steps cannot be automated. We must think about what the constraints actually imply in the stochastic setting, and we must be particularly concerned about the handling of feasibility. Very often feasibility is not as strict as we imply by our modeling, and using penalties is better.

Feasibility in this context has two components. First, constraints that seem reasonable in a deterministic setting turn into worst-case analysis in a stochastic environment. Although that might be what we want, it is rarely the case. The difficulty is simply that in a deterministic setting, where parameters normally are expected values, the constraints seem reasonable, even if they in fact represent goals that might be violated. After all, the model only handles the average case.

But in a stochastic setting, if the constraints actually represent wishes or goals and violations can be allowed, though possibly at a high cost, the constraints should be moved into the objective function with a penalty. Except for what we might call bookkeeping constraints—inventory out equals inventory in plus production minus sales—not using constraints in stochastic models should be avoided.

Constraints that remain constraints in the problem formulation are called *hard* constraints, while those moved into the objective function are called *soft* constraints. Our claim is that from a modeling perspective, most constraints are soft.

The second component of feasibility is that many, if not all, deterministic models lack a proper stage structure. Even though a deterministic model may include time periods, the variables are not properly adjusted. A side effect of this is that when the information structure of an event tree is added to a problem, the variables must be redefined. For example, in the news vendor example of Sect. 1.2, we had the same variable for orders and sales. This might be fine in a deterministic world where you never produce something you will not need, but in a stochastic model, you must realize that production and sales belong to different stages of the model and cannot be represented by the same variables. A more straightforward example was seen in the overhaul project example in Sect. 2.2, where we had to impose a scenario index on some variables. That amounts to defining new variables, even though we continued to use the old variable name.

Chapter 3

Modeling the Objective Function

> If you do not know where you are going, every road will get you nowhere.
> – *Henry Kissinger*

> A goal without a plan is just a wish.
> – *Larry Elder*

The objective function of a mathematical program is what an optimization procedure uses to select better solutions over poorer solutions. For example, if the objective is to maximize profit, then the procedure tries to move in the direction of solutions that increase profit while still remaining feasible. But when the profit depends on a parameter that is uncertain (like prices tomorrow), then the notion of maximizing profit is no longer very simple.

3.1 Distribution of Outcomes

The broadest perspective you could take on this question is that your decision, once taken today, results in a distribution of outcomes. Your choice amounts to a choice of one distribution from a whole family of outcome distributions "parameterized" by your decision. If you ignore the randomness in the problem (as many do, but of course you are not one of them!), then your procedures will still select one distribution from this family—but you will be unable to control that choice. You need some way to inform the optimization process to select distributions with favorable characteristics.

> Making a decision can be viewed as choosing one particular outcome distribution over all others.

But what is a favorable characteristic of an outcome distribution? This is a question with no simple answer. There have been many scientific avenues of

A.J. King and S.W. Wallace, *Modeling with Stochastic Programming*, Springer Series in ORFE, DOI 10.1007/978-0-387-87817-1_3, © Springer Science+Business Media New York 2012 inquiry into this issue and no one way of looking at the question has emerged, and many strange paradoxes remain. We will be content to indicate some practical concepts that can be used to select better outcome distributions over poor outcome distributions.

3.2 The Knapsack Problem, Continued

Let us first review the knapsack problem in its soft-constraint formulation:

$$\max_{x_i \in \{0,1\}} \quad \sum_{i=1}^{n} c_i x_i - d \sum_{s \in \mathcal{S}} p^s \left[\sum_{i=1}^{n} w_i^s x_i - b \right]_+. \tag{3.1}$$

Consider a solution $\hat{x}$. Unless we have a large capacity, there will be a collection of sample points $s \in \mathcal{W}(\hat{x})$ representing scenarios where the combined weight of the selected items turns out to be larger than the maximum weight allowed. So the distribution of objective function values will be a random variable:

$$V_s(\hat{x}) = \begin{cases} \sum_i (c_i - dw_i^s)\hat{x}_i + db & \text{if } s \in \mathcal{W}(\hat{x}), \\ \sum_i c_i \hat{x}_i & \text{otherwise.} \end{cases}$$

To analyze a solution to a stochastic program, you will need to examine the distribution of the objective value and ask yourself: what are the features that we are concerned about?

Perhaps the most important practical features of the distribution are the expected value and the upper and lower quantiles, for example, the 10% and 90% quantiles. Are these what you expected? Should the penalty d or the weight limit b be adjusted?

The main point we wish to make here is that it is a big challenge to design an objective function that fully captures all the desired features of the outcomes, and it is very rare to get it right the first time!

Soft constraints partition the underlying sample space into favorable and unfavorable outcomes. It is perhaps worth examining this partition to learn whether this is what was intended.

Consider a knapsack problem with two different customers. If the total weight of accepted items from the two customers were very different, this might be a bad outcome. For example, let us say that items in the set O_1 come from the first customer class and items in the set O_2 come from the second customer class. We can track this difference by calculating the expected difference in the weights of the excluded items:

$$\sum_{s \in \mathcal{W}(\hat{x})} p^s \left| \sum_{i \in O_1} w_i^s \hat{x}_i - \sum_{i \in O_2} w_i^s \hat{x}_i \right|.$$

When this difference is too large, it might be bad for customer relations! Now, think for a minute—how could you change the problem to address this issue?

Consider also the expected contribution of each item to the objective function, which can be modeled as the profit for including the item less its expected contribution to an overweight situation:

$$\left(c_i - d \sum_{s \in \mathcal{W}(\hat{x})} p^s w_i^s \right) \hat{x}_i.$$

Slicing and dicing the contributions by item attributes may lead to important insights into other features that need to be controlled. The main point is that you should look carefully at the outcome distribution and verify whether its properties were intended.

3.3 Using Expected Values

The most commonly used way of comparing two outcome distributions is to compare the expected values. By "using expected values" we do not mean that you are using a single point (the mean) as the realization of the random parameters. Rather we mean that you are optimizing the expected value of the outcome distribution.

Sometimes we choose an expected value criterion because this is the simplest and most convenient approach. But in many cases it is also the right thing to do. Here we will outline the most common arguments for maximizing expected profit or minimizing expected cost.

3.3.1 You Observe the Expected Value

Where you face a situation that will be repeated over and over again, then simple repetition favors the expected value criterion.

For example, the news vendor of Sect. 1.2 makes a fixed order that will be used daily, say, for the next year. In such a case, even if there are severe variations in daily costs, it is still reasonable to minimize *expected* cost. The law of large numbers takes over. Your annual result may be so close to the expected cost that you would not care about the difference.

> If you are planning for outcomes that will be repeated many times, then it is likely correct to use expected value as the criterion for choosing an outcome distribution.

Before rejecting the mean value as an objective criterion, you should dig into the operational details of how the decisions will be used in operations. For example, even in the case where the news vendor revises the order on a weekly basis, the average of the weekly costs over a year will also approximate the mean value—even though the variation in the weekly costs may be quite large. (How would you go about verifying this statement?)

3.3.2 The Company Has Shareholders

Making decisions under uncertainty in a corporate setting has some special features that are the subject of a deep and extensive literature. We will base our discussion here on a paper by Froot and Stein [15]. The basic idea is that a public company has shareholders who themselves are making decisions under uncertainty about how many shares they wish to hold in which company. Very often shareholders want to be exposed to risk since risk also means opportunity. The question is what sort of risk management should be pursued by the company?

Let us suppose that the company faces a major internal decision with risky outcomes. For simplicity of argument, let us also assume that even in the worst of circumstances, the company will not go broke. (The issue of bankruptcy is discussed below.)

Here are a couple of questions:

- Would you be risk averse in this situation or go for the expected value?
- Would you be willing to buy insurance to avoid the worst outcomes?

The way to answer this argument, according to [15], is to put yourself in the shoes of one of your shareholders. Let us represent the shareholder by a wise lady who understands that most of the risk in your profit is caused by your technology choices. She understands that companies may have different solutions to the problem at hand, and your success (or failure) depends on which solution your customers end up preferring.

To hedge this uncertainty concerning technologies, she buys equally many shares of your competitor's company. From her perspective, there is no longer any risk—except those that relate to her exposure to the market for your and your competitor's products.

What will happen now if both companies recognize the risks they are facing? For example, suppose they both decided to insure their technology risks? Well, our wise lady still faces no market risk, but now, whichever technology wins, she gets less! She will not be very happy with this decision.

If only one of the companies reduces risks, then it is even worse. Possibly without knowing it, she now faces risks since the symmetry between the two companies is gone *and* the expected value of her investments has gone down. So to assure investors like her that she can safely invest in your company, she must be assured that you will not behave in a way that increases the risks for her.

> A company with shareholders should maximize expected profit and leave risk adjustments to the owners, except for those risks that can be hedged more cheaply by the company itself.

Some risk-reducing measures are only available to companies, as a result of taxation rules, perhaps, or as a result of access to markets. If such measures are available to a company, then it can and should be risk averse in those respects. It is crucial that investors realize that this is being done. Many companies make statements to this respect in their public reporting. If you run into this problem in a practical setting, it is wise to consult with financially trained people. Our point is that you become aware that even in decisions where you do not observe repeated outcomes, it may not be appropriate to reduce all the risk you face.

Our wise lady's strategy is related to a very popular statistical arbitrage called "pairs trading," in which investors take a short position in one competitor and an offsetting long position in another. The investor is not exposed to the success or failure of the underlying market but will make money on a temporary deviation that supposedly should revert to the statistical long-run mean. This kind of gamble is really a byproduct of how large corporations are managed. There really is no way to reward management for outcomes based on whether a given technology is good or bad. The actual practice is to reward *steady earnings growth* of the sort that can only be achieved by holding a diversified portfolio generating the products and services sold by their sales forces. If two large companies hold diversified portfolios and comparable brands, then taking bets on statistical properties of their earnings is a plausible arbitrage strategy.

3.3.3 The Project Is Small Relative to the Total Wealth of a Company or Person

Even if the variance of the income from an investment is very high relative to the mean, this variability is usually not cause for concern if the numbers happen to be small. When someone tells you: "The cost will be two or three dollars, I am not sure which," you probably say: "OK, I don't mind." You are not concerned despite the fact that one estimate is 50% higher than the other.

On the other hand, if the same person tells you, "Oh, it will cost two or three million dollars," you probably would hesitate, even if the ratio between the two cost estimates is the same. Why? Probably because in the latter case, the amounts are substantial relative to your total wealth.

> Use expected value if the relative importance of the project is small.

This is actually a variant of the first argument about observing the expected value. If a project is small relative to a company's total wealth, then the company probably has a very large number of these projects, and it will observe the expected value (but now the expectation is over projects and not over outcomes within a single project).

If none of these arguments applies, then it is probably time for you to think properly about what risk actually means in your case and if you should be worried about it.

3.4 Penalties, Targets, Shortfall, Options, and Recourse

When soft constraints are incorporated into an objective function, the part of the objective function that models the soft constraints will have a certain shape. Again, look at the knapsack problem (3.1). The soft constraint part of the objective is the expected value of a piecewise linear function of the excess weight. It is zero below the capacity b and has slope d above the capacity. This piecewise linear function has many names and appears in many contexts.

3.4.1 Penalty Functions

In the language of linear programming, such functions are sometimes called *penalty functions*. The penalty function has two parts. One is the *target interval*, which in the knapsack case is the interval below the capacity, namely $(-\infty, b]$. The second part is the *penalty rate*, namely the rate at which the penalty accumulates as the target is missed. This rate may be an actual cost of responding to the penalty, but more usually it is a modeling approach that is used by the modeler to shape the outcome distribution. An example for the knapsack problem can be found in Fig. 3.1.

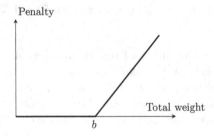

Fig. 3.1: Penalty function for knapsack problem

3.4.2 Targets and Shortfall

One can use penalty functions to indicate a desire to reach a target. Although there is a similarity to a penalty formulation of a soft constraint, a

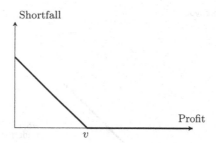

Fig. 3.2: Example of shortfall function with a target v

target formulation is not really a soft constraint since the act of selecting a target reshapes the outcome distribution. If you prefer outcomes x^s above a given target v to outcomes below the target, then the simplest criterion that measures this preference is called the *shortfall measure*:

$$\sum_{s\in S} p^s(v - x^s)^+. \tag{3.2}$$

Consider the function

$$h(y) = \begin{cases} 0 & \text{if } y < 0, \\ y & \text{if } y \geq 0. \end{cases} \tag{3.3}$$

This is a piecewise linear function with slope 0 below zero and slope 1 above zero. An illustration of a penalty function can be found in Fig. 3.2. It is not hard to see that

$$\sum_{s\in S} p^s h(v - x^s) = \sum_{s\in S} p^s(v - x^s)^+. \tag{3.4}$$

This function compares two outcome distributions by looking at the expected values over the region below the target v. So in the situation where x^s represents the return of an investment portfolio over some time horizon, then perhaps we would like to avoid outcomes with higher values of $h(v-x^s)$. If we are in a maximizing frame of mind (most investors are), then we could maximize the following expression:

$$\max_{x\in X} \left\{ \sum_{s\in S} p^s x^s - \sum_{s\in S} p^s h(v - x^s) \right\}, \tag{3.5}$$

where X is some set that constrains our portfolio choices. What are we maximizing here? We are maximizing a piecewise linear function, which, computationally, is not too hard. What would it do? Well, if we had two outcome distributions with similar means, this method of choosing outcomes would prefer the one with a larger *conditional* expectation over the outcomes that lie above the target v.

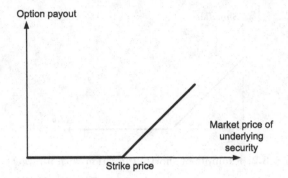

Fig. 3.3: Value of call option as a function of the price of the underlying security

These types of shortfall measures are very useful in applications involving uncertainty. Target shortfall measures are a natural way to describe differences in outcome distributions, and the optimization technology required to solve them is readily available. We will find many examples of these in this book.

3.4.3 Options

Penalty functions model a cost that is incurred when certain underlying events occur. In finance, a contract with terms that incur a cost or produce a payment depending on the occurrence of a future event is called an *option*. A typical type of option is a call option, which grants the owner the right to buy an underlying security at a fixed price, called the strike price, at some fixed date in the future.

The owner of the call option has a choice on the exercise date. If the price of the underlying is above the strike price, then the owner can buy the underlying security for the strike price and then sell it at the market price. The owner's profit equals the difference between the market price and the strike price. On the other hand, if the market price is below the strike, then the owner need not do anything. The option payout is a function that is zero below the strike and increases linearly with slope 1 above the strike, as in Fig. 3.3.

Can you see that an option payout (Fig. 3.3) looks like a penalty function (Fig. 3.1) with a target equal to the interval above the strike price and a rate equal to one? Also note the similarity with the shortfall function (Fig. 3.2). Many penalty formulations can be framed in terms of call and put options because penalties are essentially invoked when a stochastic value goes above or below a target.

3.4.4 Recourse

Another term that applies to penalty formulations is *recourse*. A recourse model describes actions that lead to a future cost or benefit in response to future events. In the case of a penalty formulation, the recourse model just calculates the penalty (such models are called "simple recourse" in the literature). But a recourse action could be more complex. A recourse model can minimize the impact of a bad event using multiple technologies that are available to the decision maker but that may not be available to investors.

Investors use options to implement strategies for portfolio management. In the same way, decision makers may invest in recourse capabilities to improve their capabilities for managing uncertainty. In a sense, a stochastic program with recourse can be viewed as an option portfolio selection model. However, recourse is a concept that goes far beyond options. Recourse is modeled from the collection of possible actions and resources available to the decision maker, so in a sense, the use of recourse models allows decision makers to design their own options.

3.4.5 Multiple Outcomes

Typically in stochastic programming, you are concerned with multiple outcomes. In the foregoing example, we were concerned with maximizing the mean return and at the same time wanted to minimize the shortfall measure. The natural thing to do was to parameterize the problem:

$$\max_{x \in X} \left\{ \sum_{s \in S} p^s x^s - \lambda \sum_{s \in S} p^s h(v - x^s). \right\} \qquad (3.6)$$

Varying the parameter gives an "efficient frontier" of solutions; $\lambda = 0$ gives the solution that maximizes expected return, and as λ grows higher and higher, it will give the solution that minimizes the shortfall measure.

This "multiobjective" style of optimization procedure is very common in stochastic programming. It allows us to describe multiple targets and objectives, and the optimization process generates efficient sets of solutions, each with different properties relative to the targets.

3.5 Expected Utility

The problem of choosing outcome probability distributions has a very deep technical literature that centers around the economic utility theory developed by von Neumann and Morgenstern [54]. The basic idea is that preferences between outcome distributions (following certain rules) can be modeled by choosing an outcome distribution that maximizes *expected utility*, where utility is modeled by a concave function of the outcomes. We will give just a brief outline here of the portfolio selection problem in finance since it is in finance that the basic assumptions of expected utility are likely to be satisfied.

Let us place ourselves in the realistic world of choosing to invest in corporations that are in effect managing portfolios of businesses. As observed previously, one can anticipate some statistical regularity of outcomes that will be observed as dividend payments or changes in the market prices of company stock. Purchasing a single share of stock in company i will produce an annual return of $r_i(s)$, where s is a scenario parameter indicating the strength of the market returns modified by the idiosyncratic performance of management. (Just to be clear on the meaning of *return*, we will adopt the convention that a return less than 1.0 represents a loss and one greater than 1.0 a gain.)

Investing in a portfolio $x = (x_i, i \in I)$ of companies will therefore produce an outcome distribution of

$$s \mapsto \sum_{i \in I} x_i r_i(s), \text{ with probability } p(s). \tag{3.7}$$

Since companies are run by managers who are rewarded for earnings growth, it is quite likely that there is some statistical regularity to the dividend payments. Under the assumptions of *expected utility*, then, there exists a utility function $F(\cdot)$ such that the optimal choice of outcome distribution is given by maximizing expected utility:

$$\max_x \sum_{s \in S} p^s F\left(\sum_{i \in I} x_i r_i(s) \right). \tag{3.8}$$

Now we ask the question—what should the utility function be? In the case of expected value optimization, the utility function is just the identity $F(R) = R$. Should we use the expected value criterion to choose an optimal collection of stocks? What are the other choices?

If our perspective is very long term (for the rest of our long lives, for example) and our objective is to simply take the money every year and *spend it*, then the expected value discussion applies: the variability over many many years will oscillate around the mean.

On the other hand, if we take the money and *reinvest it*, then the story is different. When the returns are identically distributed, the strategy that achieves the maximum wealth is the one that *maximizes the logarithm* of the return. (Of course, this is simply the mean of the exponential growth one achieves through reinvestment—so the mean wins out here, too!). This result is originally due to Kelly and has been developed in the stochastic programming context by Ziemba and his colleagues [41].

Of course, these objectives assume we know the distributions rather precisely. In fact, the distributional characteristics of investment returns change over time. The next section investigates an important tool used by portfolio managers to model the risks of investment.

3.5.1 Markowitz Mean-Variance Efficient Frontier

You have likely heard of the Markowitz criterion, in which the objective minimizes the variance for a given level of expected return [44]. Why is this so popular? Well, for one thing it uses observable statistics—the mean and variance of financial returns are easily observable. But is it sensible? After all, variance penalizes both downside risk and upside risk. Is it reasonable for an investor to choose a criterion that minimizes the risk of going higher?

To answer this question, let us consider an investor who knows her statistics. For instance, she knows the mean return vector m and the variance-covariance matrix V. She also knows her von Neumann–Morgenstern and wants to choose her portfolio according to maximum utility. But what utility function? She goes to a Web site that offers to discover her utility by asking questions about one gamble after another. But she is really not sure about this at all. So many comparisons!

Along comes a slick stochastic programmer who offers to give her an *entire collection of optimal portfolios* to choose from. Every one will be optimal for some utility function. And up to a second-order approximation, every utility function will be represented. How does he do it?

He argues like this. Suppose your utility function was $F()$ and we knew how to find its optimal portfolio, namely $\hat{x}$ maximizes (3.8). Then of course we know its expected return, namely $\hat{R} = \sum_i m_i \hat{x}_i$. Expand the utility function to second order around this expected return:

$$F(R) \sim F(\hat{R}) + F'(\hat{R})(R - \hat{R}) + 1/2F''(\hat{R})(R - \hat{R})^2. \qquad (3.9)$$

Now find the maximum utility using the right side of the approximation instead of the left:

$$\max \sum_s p^s \left[F(\hat{R}) + F'(\hat{R}) \left(\sum_i r_i(s)x_i - \hat{R} \right) \right.$$
$$\left. + 1/2F''(\hat{R}) \left(\sum_i r_i(s)x_i - \hat{R} \right)^2 \right]. \qquad (3.10)$$

Without loss of generality, let us also restrict the search to those choices that satisfy

$$\hat{R} = \sum_i m_i x_i. \qquad (3.11)$$

This does not specify the choice of x (it does narrow it down considerably, but let us keep going). The main point to keep in mind in the argument is that the choice of F determines $\hat{R}$. Now with this narrowing down, let us look carefully at the approximate utility maximization. First, note that the term $F(\hat{R})$ is fixed, so it will be ignored in the approximate maximization. Second, note that the second term disappears! This is because we are restricting our choices of x to those that lie on the mean hyperplane (3.11). Finally, note that the last term consists of the second derivative of a concave function (which is negative) multiplied by the variance of the return. When we clear the negative term from the objective, the maximization turns into a minimization and we are left with the problem of minimizing the variance subject to a constraint on the mean return.

It follows that the approximation is none other than a version of the mean-variance problem central to the Markowitz method:

$$\min \quad \sum_{i,j} x_i V_{ij} x_j$$
$$\text{such that } \hat{R} = \sum_i m_i x_i. \tag{3.12}$$

As we vary our choices of utility $F()$ we will also vary our choices of return $\hat{R}$. It follows that all our choices will lie on the *efficient frontier* of solutions that minimizes variance for a given level of mean return. This is the approach originally formulated by Markowitz [44]. The mean-variance efficient frontier does in fact present our investor with a collection of points that a utility-maximizing investor would choose, up to a second-order approximation.

How good is the approximation? Well, this is something you can try for yourself. Find some tables of annual returns of large corporations over the past 20 years, calculate the means and variances, and answer for yourself: how good is the second-order approximation to your favorite utility function, say, the logarithm? You will find that it is pretty close. After developing this argument (in [34]) we asked this question. For the logarithm function it seemed like the second-order approximation was very sensible for absolute returns in the range of 75–400%—which is one very good reason for the popularity of the Markowitz method over the 60 years since its discovery.

The other reasons are that the parameters are quite easily observed in the marketplace. The covariance matrix and mean can be constructed by observing a time series of annual returns. The sequence of observations is viewed as samples drawn from the distribution of future returns. Standard statistical calculations can be used to provide appropriate estimates. It appears that there are long-term cycles in *market* volatility; however, the same cannot be said about variances and correlation terms for individual stocks—partly because there is less data to estimate them and partly because the relative performance of company stock prices depends on so many factors. However, the actual performance of the mean-variance model over time is much more sensitive to the estimation error in the fundamental parameters, most especially the mean.

At this point we will close this discussion. This is not a book about statistical arbitrage in financial markets; it is a book about modeling choices under uncertainty. We hope we have conveyed to you some of the flavor and language of expected utility and the mean-variance approach as it is applied in investing.

The interested reader can go much further, of course. However, in the end, we would like you to be aware that many market practioners, on the basis of much experience, do not believe that past data inform the future behavior of prices. Rather, the basic reason for prices to move one way or another is due to supply and demand—and in the securities market the dynamics of supply

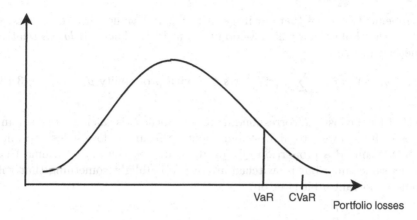

Fig. 3.4: Density function, VaR, and CVaR for potential losses. VaR corresponds to a chosen quantile, while CVaR is the expected loss above VaR

and demand are affected at times by overwhelming optimism and at times by overwhelming pessimism and always amplified by leverage. This brings us to our next topic.

3.6 Extreme Events

The 50-50-90 rule: Any time you have a 50-50 chance of getting something right, there's a 90% probability you'll get it wrong.
– *Andy Rooney*

Sometimes extremely bad things happen. Asteroids strike, virulent diseases break out, markets crash, products fail. Models with uncertainty must consider the consequences of extreme events.

In the financial industry, for example, regulators require some institutions to estimate the upper tail of the loss distribution and to hold reserves proportional to these loss estimates. The intention of the regulators is to force the industry to model the worst-case extreme losses of their portfolios and to penalize extremely risky positions by forcing them to hold safe, but low-yielding, securities in proportion to the extreme-loss potential. The upper quantile of potential losses is called *Value at Risk* or VaR. A related, increasingly popular, and, in our view, preferable statistic is called the *Conditional Value at Risk* or CVaR—which is the expected value of the losses that fall above the VaR. This is illustrated in Fig. 3.4.

The first step in analyzing extreme risks is to decide what to analyze. In the preceding investment problem, we could decided to set a *benchmark* return. For example, we could have chosen a level R_s of returns (keep in mind that a return below 1.0 means that our investment has lost value) that represents an absolute threshold below which we do not want to go. Then our *losses* relative to the benchmark are

$$L^s(x) = R_s - \sum_i x_i r_i^s \quad \text{for } s \in \mathcal{S} \text{ with probability } p_s. \tag{3.13}$$

The VaR for the losses L corresponding to a quantile Q is calculated by sorting the losses from lowest to highest and counting from the lowest loss up until you have counted a proportion $Q\%$ of all scenarios. Here is a formula that says the same thing but is switched around to highlight something that will become important a bit later on:

$$\text{VaR}(x; Q) = \inf_L \text{ such that } \#\left\{ s \in \mathcal{S} \text{ with } L^s(x) - L \ge 0 \right\} \ge \frac{1 - Q}{100}, \tag{3.14}$$

where the number sign $\#$ in front of the set indicates the number of points in the set. The risk-level $\text{VaR}(x; Q)$ divides the sample space of losses into two parts: good and bad. The good points are the $Q\%$ that have losses below the $\text{VaR}(x; Q)$ and the bad points are the $(100–Q)\%$ that have losses greater than the $\text{VaR}(x; Q)$.

In banking and finance one says "our 99% VaR is USD 0.5B," and it is understood by bank regulators that the probability that the bank's position will suffer a loss greater than $500 million is only 1%. These statements are taken quite seriously. In fact, the bank must demonstrate to the regulators, by reconstructing their positions backward in time, that their past 200 daily 99% VaR loss estimates were violated no more than twice.

But can you see that VaR behaves a bit strangely as a function of x? Try this thought experiment: take a position x_i in the security with the absolute worst losses and make a small increase in it. A good risk measure should increase when you do more bad things. Is the VaR guaranteed to change? No, it is not, and this is why it is a controversial measure.

The CVaR is defined as the expected value over the quantile in definition (3.14) of VaR. Given that expected values are linear, one would anticipate that a risk measure based on CVaR will behave as a risk measure should. However, there is the complicating matter of locating the support of this expected value, which behaves badly, as we have seen. A relatively new result by Rockafellar and Uryasev shows that CVaR is indeed a well-behaved risk measure. They proved that its value is given by the following stochastic program:

$$\text{CVaR}(x; Q) = \inf_L \left\{ L + \frac{1}{1 - Q} \sum_s p^s \max[0, L^s(x) - L] \right\}. \tag{3.15}$$

Do you see that this calculates expected value over the same set that we counted points over for the calculation of VaR?

In usage, VaR and CVaR have different purposes. Both are measures of the tail behavior of an outcome distribution. VaR tells you something about where the tail is supported, whereas CVaR gives you more information about how the distribution behaves over the tail.

Regulators prefer VaR because they have more confidence in the statement "only 5% of losses are greater than VaR" than they do in the statement "the average of the worst 5% of losses is CVaR." The second statement requires a model of the tail distribution, which by necessity must always be based on a sample of very few observations. People can disagree about tail distributions, but the definition of the location of the tail is usually pretty well estimated. Regulators will prefer to stick with estimates for which there is broad agreement.

On the other hand, decision makers should prefer CVaR, for two reasons. First, precisely because it does require a model of the tail distribution, the inclusion of CVaR will force decision makers to think about how bad things can get.

We can think of CVaR as being just like a target shortfall measurement, but where the target is specified in terms of the quantile instead of some arbitrary target. Of course we could just specify the target to be high enough to be "in the tail." The advantage of using CVaR is just the fact that we do not have to guess beforehand where these tail values are.

Sometimes it is a good idea to use a different distribution for extreme events. For example, on trading floors, the traders do not use the same distributions as the risk managers. Risk managers have different objectives and different time horizons. A trader about to make a quick killing should not be concerned about surviving a low-probability market crash over the next 30 days. It is the risk manager's job to worry about that. The risk manager does not do this by influencing the trader's models. (It is hard to imagine a trader accepting any kind of modeling advice from a risk manager!) Risk managers do their job by limiting trader access to the firm's capital. It is entirely reasonable that the risk manager and the trader use different distributions to achieve their objectives.

Extreme event modeling in finance is gradually being extended by regulators to cover so-called operational risks. This covers things like power failures at data centers, losses due to failures of internal controls, rogue traders (and risk managers), and so forth. The financial system is so interlinked and the traded volumes so great that an operation breakdown, say, due to a power failure at a major bank's data center, could have widespread financial and economic consequences for weeks afterward.

Companies in industrial sectors with large extended supply chains are also beginning to model the tail distributions of bad events, operational and otherwise. Examples of bad events could be the bankruptcy of a key supplier or creditor, quality control failures internally or of key suppliers, a major product liability lawsuit, improper release of private information of consumers, earthquake damage to a key electric power supplier, or losses in financial portfolios that back employee insurance programs or working capital.

3.7 Learning and Luck

I'm a great believer in luck, and I find the harder I work, the more I have of it.
– *Thomas Jefferson*

Learning is a subject of its own, and we will not have a general discussion of the issue here. However, since such things as learning organizations and organizational memory are preached by many consultants, we would like to mention but one issue that is relevant for this book. Learning presumably implies that from the outcome of a decision we wish to become wiser. Next time we make a similar decision we wish to either make a better decision (if this one was not so good) or an equally good one (if this one was good). For this to make any sense, there must be a causal relationship between what we did and what happened.

If you walked backward into the store where you bought the winning lottery ticket, you might think (many certainly do) that this is good for winning, and you might choose to do the same in the future. We all know that this is nonsense, and we call it superstition, not learning. If, instead, you have a model you use for making some decision, and the decision leads to, say, a very good outcome, do you then know it was a good decision and a good model? Is there a logical connection between the ex-post observation and what you did? Let us again exaggerate a bit. Assume there are two gambles, both with the same price for taking part and both with the same winning prize. Assume in one gamble there is a 10% chance of winning the prize, in the other 90%. Assume you play the game with a 10% chance and, voila, you win! Does that mean you made a good decision? Of course not. Is there anything to learn from the ex-post observation? No, there is not. You were simply stupid and lucky. We can say that because there is an ex-ante evaluation that in this simple case tells us that what you did was stupid. The fact that you were lucky does not change that.

So what is learning? We will not answer that question. But be aware that learning is a difficult issue in a random environment. In genuine decision contexts, there is a causal but stochastic relationship between what you do and the consequences. But learning can be hard because you must separate luck from cleverness.

But do we care? Maybe for our own decisions we often do not. We prefer to value our decisions based on ex-ante analysis, looking for decisions that according to our own utility function maximizes expected utility. But there is a context where we really are interested. If you are to engage a consultant, he probably sells himself based on his track record (sounds great, does it not?). Here we are interested in ex-post learning. Is he really good or just lucky? Maybe he looks so good because he takes too many chances, and for that reason he is definitely not the one we want. Some say that if your broker makes a lot of money for you over a reasonable period of time, fire him! He takes too big risks.

Chapter 4

Scenario-Tree Generation: With Michal Kaut

It is a very sad thing that nowadays there is so little useless information.
– *Oscar Wilde*

Weather forecast for tonight: dark.
– *George Carlin*

So far we have talked about models and structures with respect to uncertainty. But we have not talked about data. We have talked about prices and demand, deterministic or stochastic, we have talked about distributions of random variables. But rarely are these available in the format we need for our algorithm.

We shall not bring up most questions of data and data manipulations and estimations here. Much of that belongs elsewhere, in your course in statistics, econometrics, or maybe forecasting. Maybe your data comes from experts, or maybe all you have is your own humble understanding of the random phenomena governing your model. Whatever you have, this book picks up the modeling process at the point where you have all the data you are able (or willing) to collect. So you may have historical data, possibly vast series of them, you may have actual distributions, estimated or assumed.

In order to find a solution to your stochastic program you will most likely need to create a *discrete version* of your probability distribution. Many algorithms used to solve decision problems under uncertainty require a discrete distribution, especially algorithms designed for general purpose problems. (Some problems are designed, with specialized distributions and compatible formulations, to be solved "in closed form" without a discretization step.)

It is important to realize from the very beginning that the resulting discrete distribution is not part of the actual problem under investigation, but something you might need *because of the algorithm you have chosen*. When investigating your problem, your interest is most likely the random variables themselves (dependent/independent, normal/log normal/uniform, . . .) and of

A.J. King and S.W. Wallace, *Modeling with Stochastic Programming*,
Springer Series in ORFE, DOI 10.1007/978-0-387-87817-1_4,
© Springer Science+Business Media New York 2012

course the algebraic formulation of the model. But in most cases such a model cannot be solved as a stochastic program. Hence, you need to discretize. However, since there is no obvious way to pass from random variables to a (good) discretization, we choose to see the discretization as part of the modeling. Different people will discretize the same data in different ways, and hence, we can see the discretization as a way to model the random variables. Of course, it is also valid to view the discretization as part of the solution procedure. We have not chosen that view, however, mostly because for most algorithms used to solve stochastic programming problems, there is no obvious way to go about the discretization, and hence, we wish to point out that the process must be governed by the modeler, that is, you.

> It is your responsibility to make sure that random variables are represented in an appropriate way. There is no general way that always works.

Because the discretization is not part of the problem, it may not interest you from an academic point of view. But how you do the discretization is actually crucial for your results. Hence, this chapter has two purposes, one is to explain *why* this part of your modeling process is crucial, and another how you can try to go about the discretization at minimum effort if this is not your real area of interest.

At the simplest level, the goal of the discretization procedure is to make sure that it is not the discretization procedure itself that drives your optimization problem, but rather your model—meaning your algebraic formulation and your model of the random variables. In a sense, the goal is to totally hide the discretization procedure so that things are *as if* you had solved the model with the original distributions. If there is a user of your model, that user is most likely interested in the algebraic model (or at least what it represents) and may be interested in discussing the random variables as such. But he or she will hardly ever be interested in how you discretized the distributions. They will tell you that "that's your problem".

> A good discretization procedure cannot be observed in the solution to the model.

Let us be a bit more specific about why this is important. Assume that you, for example, have a stochastic network design problem, and you wish to find out how the optimal design depends on the expected demand between nodes a and b. Hence, you solve your stochastic program several times, varying the expected demand over the relevant area. You obtain a certain set of solutions, and you study the solutions to understand how the design depends on the expected demand. But your discretization procedure also depends on the expected demand. As the expected demand varies, you will be producing different discretizations, and these are used in the optimizations. Your concern now ought to be: Are the observed changes in optimal design a result of the discretization, or are they actual changes in the optimal design? Are the observed changes real, or are they systematic or random effects of how you discretized? Are you sure you now know how the optimal design depends on the expected demand between nodes a and b? Or is everything just noise from your discretization? The purpose of this chapter is to help you ensure that it is not the discretization that drives the results.

The premise in this chapter is that you have chosen a solution algorithm for your stochastic program that simply requires a scenario tree as input. However, be aware that there are some cases where this is not needed.

First, there are special *solution algorithms* that have sampling included as a part of the solution procedure. One example is stochastic decomposition, as defined by [24, 25]. Also this algorithm uses scenario trees, but they are sampled as you go along. The algorithm does not suffer from the difficulties that will be mentioned in Sect. 4.1.1. Be aware that the algorithm is very challenging to implement due to its dependence on efficient data structures, and that it can be used only for stochastic *linear* programs.

There are two other such methods that work for convex, not only linear, problems: stochastic quasi-gradients by Ermoliev and Gaivoronski [13, 16], and importance sampling within Benders decomposition by Dantzig and Infanger [8, 29].

There are also special cases, such as versions of the newsboy problem (discussed in Chap. 6), where specialized algorithms have been developed for special cases, such as normal distributions, and where discretizations are not needed.

4.1 Creating Scenario Trees

We shall, as we mostly do in this book, present the theory in a two-stage setting. Hence, there will be no time index on the random variables. In multistage problems, discrete random variables will, due to dependence, take the form of a tree, where the root represents "today" and the tree branches for each stage. In the two-stage case, the tree is simply a "bush", with one node for "today" and one node for each possible "tomorrow". Even so, we shall allow ourselves to talk about scenario trees also in that case.

The goal is now to create a scenario tree such that the error caused by discretization is as small as possible, while maintaining solvability of the model.

4.1.1 Plain Sampling

An obvious way to create a scenario tree is to sample from the underlying distribution. If you do this, eventually, you will end up with a tree (discrete distribution) that is arbitrarily close to the distribution you started out with. So why is this not the whole story? Why not simply sample a "large enough" tree, and then proceed? The answer is simple enough. For all but the simplest problems, you will be left in one out of two situations:

1. Your tree is so large that the discretization error is acceptably small. But it is also so large that you cannot solve the optimization problem. You have a numerical problem.
2. The tree is so small that you can solve the optimization problem, but also so small that the discretization is a bad representation of the underlying distribution. You have a representation problem.

There is a theoretical basis for this statement as well as practical experience. Shapiro [49, 50] shows that

$$N \geq \frac{12\,\sigma_{\max}^2}{(\epsilon - \delta)^2} \left(n \log \frac{2\,D\,L}{\epsilon - \delta} - \log \alpha \right),$$

which "shows that the sample size which is required to solve the true problem with probability $1 - \alpha$ and accuracy $\epsilon > 0$ by solving the sample average approximation (SAA) problem with accuracy $\delta < \epsilon$, grows linearly in dimension n of the decision vector x" [50, pp. 375]. Here, D is diameter of the feasible region, L is the Lipschitz constant of the objective and $\sigma_{\max}^2$ is a bound on variance of a function of the objective, used to derive the result.

This might look daunting, and indeed it is. If you like to try and put in reasonable numbers, you'll see that N has to be extremely high for reasonably sized problems. So if you wish to guarantee a reasonable quality solution for a reasonably large problem, you are indeed in trouble.

Sampling might of course in many cases lead to much better solutions than indicated by this inequality. But to realize that has happened you need other tools—tools that test the quality of a given solution. We shall return to that later.

So, there is an important point here. A procedure that has all the required properties in the limit, may not have any desired properties if not taken to the limit. A tree made by a procedure that has perfect limit properties may still be arbitrarily bad for its purpose.

> Although sampling has perfect limit properties, there is no guarantee that a small sampled tree has any useful properties when used in a stochastic programming model.

Sampling can be improved by correcting means and variances ex post by using a simple linear transformation on each margin. Since many stochastic programs are sensitive to expected values (see, for example, [33]), this can improve the scenarios dramatically.

Another, possibly less obvious issue, is that to sample you need to know what you should sample from. Often, you do not fully know your distribution. To sample, you might need to add information (that possibly is arbitrary) to have something to sample from. Some of the ideas to be presented in this chapter may not need the full distribution. In some cases that is an advantage.

4.1.2 Empirical Distribution

If all you have is history, and you have no qualified reason to believe that the future will be different from the past, then what do you do? In particular, do you use the data (history) to estimate a continuous distribution as part of your modeling, or is that not a good idea? Remember that, eventually, you are going to revert to a discrete distribution.

If you use the data to estimate some parameters of a distribution, you are adding information to the data. If this information is arbitrary, you are adding meaningless information. That is not a wise thing to do. On the other hand, if you know, or have reasons to believe, that the data comes from a specific family of distributions, estimating that distribution is a way to add your extra information.

So, the story is: if you have nothing but historical data, use that data as your distribution. So when we in what follows refer to a distribution that needs to be dicretized, we might be referring to the empirical distribution, as it might have too many outcomes for the stochastic program to solve in reasonable time.

You may have your own ideas about how to make a scenario tree. If so, you should certainly investigate them. But do not use them blindly. Rather, use the methods that follow in this chapter to test your methods to see if they actually behave well. Otherwise, you may find yourself in trouble later on.

4.1.3 What Is a Good Discretization?

It is worthwhile thinking for just a short while about we should mean by a good discretization. We already mentioned that it should be one that does not affect the solution to your problem. It should function in such as way that it is as if you had used the original distribution. Consider the two discrete distributions in Fig. 4.1.

In the figure, we have also drawn the density of the standard normal distribution N(0,1) for comparison. Would you say the discrete distributions are good approximations of the normal distribution? And similarly, would you say the left discretization is a good approximation of the right one?

Well, that is in fact hard to answer, and the reason is at the core of the problem of discretizations: it depends on the use! The Markowitz model for portfolio management (see Sect. 3.5.1) will give the same solution with these three distributions. In fact, in any dimension, the model only reacts to the first two moments (including cross moments) of the distribution, and the three

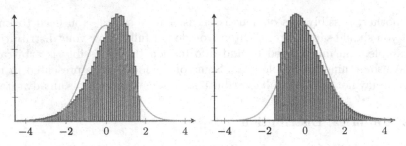

Fig. 4.1: Histograms of two discrete distributions, compared to the density of a normal distribution with the same mean and variance $(N(0,1))$

distributions in the figure have the same first two moments. So despite the fact that they look different, and in fact are very different (in particular the third moments are very different), they each serve us equally well.

> The quality of a discetization depends on the model in which it is used.

Do you think this is obvious? Well, maybe it should be. But many approaches in scenario tree generation seek a discretization which makes the scenario tree as similar to the original distribution as possible, *without regard to the stochastic program needing them*. This has the advantage that generating scenario trees can be done without knowing the corresponding optimization model, and in fact can be done by someone with pure statistics expertise. In such a setting, the three distributions in Fig. 4.1 would probably be declared rather different, and in particular, not good approximations of each other.

Now, if you find a scenario tree that is close to the original distribution in this purely statistical sense (statisticians have ways of measuring the distance between two distributions), and that the discretization leads to a solvable stochastic program, *and* the resulting model is stable in a way we shall soon describe, then all is fine. But if you have a reasonably large problem, you might be in trouble. You might find

1. That the optimization problem is not solvable, or
2. That the solution you have found is no good, even if your discretization was as close as possible to the original distribution, given the number of scenarios you requested... and if you increased the number of scenarios in the discretization to get better solutions, you found yourself in item 1 instead.

Is there a way out of this loop? And does this have anything to do with how you measure the quality of your discretization? The answer is "maybe" on the first question, and "yes" on the second.

The reason is simple enough. The more aspects / properties of a distribution you care about (moments, extreme values, co-variation ...) the more scenarios you need in order to have a good discretization. But if some of these properties, such as the third moments in Fig. 4.1 in case of the Markowitz model, do not matter to the optimization problem, why do you care about them? Think about a multidimensional distribution as a map. If you want a discretized representation of the map to look like the map itself, you may need many scenarios. But why care about parts of the map where you will not go? That is, why care about parts of the distribution that the optimization problem does not care about?

So, by only caring about the parts of the distribution that matters to your problem, you can get away with fewer scenarios in your discretization *or* obtain a better solution to your problem (i.e., get a better discretization) for a given number of scenarios, if you let your optimization problem guide you.

> In stochastic programming, the quality of a discretization is determined by the optimization problem using it.

Apart from simple cases like the Markowitz model, where the results can be established analytically, we are not implying that it is easy to find out what properties of a distribution are important for your case. So far we have only established that by *thinking* that way, you may indeed bring a problem from being not solvable to being solvable. Just looking at the distributions, without reference to the optimization problem, may be easy, but may also be fatal. We shall return to this issue.

So let us turn to the next section where we ask: Is my procedure for generation scenario trees good? That is, do I have a good discretization? This section is particularly important if you have your own ideas about scenario tree generation. Make sure you do not use your procedure without testing it along these lines.

4.2 Stability Testing

So now we are back to our initial worry in this chapter: Can we be sure that when we test our decision model, we are not just testing the scenario generation procedure? Or more extremely, could it be that the results of the optimization model are not in any logical way a result of our algebraic model and random variables, but just a random or systematic side effects of a bad scenario generation procedure? The ideas here stem from [31].

It is impossible to get all the way to the bottom of this problem. But we'll see that we can get at least somewhere. We shall discuss two issues, *in-sample stability* and *out-of-sample stability*. The first represents a test of the internal consistency of a model (remember again, we view the generation of scenario trees as part of the model) the second is related also to model quality.

We shall also here discuss the issues in terms of a two-stage model. Hence, a scenario tree is simply a bush, and we shall refer to it as $\mathcal{T}$. The problem at hand can thus be written as

$$\min_x f(x; \mathcal{T}), \tag{4.1}$$

where we have hidden the second-stage variables $y(\mathcal{T})$, and where it is implicit that we are to take expectations with respect to the second argument, that is, the discrete distribution described by the tree $\mathcal{T}$. The true problem, the one we by assumption cannot solve, is given by

$$\min_x f(x; \boldsymbol{\xi}). \tag{4.2}$$

4.2.1 In-sample Stability

In-sample stability does not have a deterministic counterpart. Assume you have a scenario generation procedure that is in itself random, that is, it does not produce the same scenario tree each time it is run with exactly the same data. This is typical of any sampling based procedure, but true for many others as well. If your procedure generates the same tree all the time, there is a corresponding discussion that we shall return to below.

Assume now that you run your scenario generation procedure several times on the same data, producing many different trees. Let us denote them $\mathcal{T}_i$. Then run your optimization model equally many times, one time with each tree, that is, solve "$\min_x f(x; \mathcal{T}_i)$". Let the optimal solutions be given as $\hat{x}_i$. If the optimal objective function values are (about) the same in all cases, that is, if

$$f(\hat{x}_i; \mathcal{T}_i) \approx f(\hat{x}_j; \mathcal{T}_j)$$

you have in-sample stability. In-sample stability is in fact a totally instant-dependent property, but from a practical point of view, we shall assume that it is valid also for closely related instances. So, if you have in-sample stability, it does not matter which scenario tree you use, and hence, when solving your stochastic program, you simply take your algebraic model, your distributions, and then run first the scenario generation procedure, and then the optimization problem. And you do it, assured that the objective function value would be the same if you ran your problem again with the same data: what you do is internally consistent.

4.2.1.1 The Scenario Generation Procedure Is Deterministic

If you have a deterministic scenario generation procedure, we suggest you instead run your procedure repeatedly asking for trees of slightly different sizes. It is unlikely that you have a deterministic procedure without any control over the size of the resulting tree. To have in-sample stability, you should observe (almost) the same optimal objective function value all the time. If you get totally different results just by adding, say, one scenario, clearly there is something wrong with your modeling.

4.2.1.2 Why Measure the Objective Function Value?

Stochastic programs tend to have flat objective functions. That is, in fact, one of the reasons for using stochastic programming. Hence, very different solutions can be more or less equal in terms of profit, cost, or whatever is the measure of goodness. As we do not want to declare it a problem to end up with a different solution if it serves us equally well as another one (after all, we are testing the scenario tree generation procedure), we choose to measure similarity by the objective function value. To achieve stability in terms of solutions is going to be very hard for problems with very flat objective functions. For example, in portfolio management, if there are two reasonably different portfolios which behave the same way, that is, which provide the same trade-off between risk and reward, we do not want to declare it a problem that they look different, at least not when the issue is the scenario tree. You are of course free to disagree with this view and require stability in solutions. Be aware, however, that will be very hard to achieve.

4.2.2 Out-of-Sample Stability

In-sample stability was a test of the model itself. It represented a kind of robustness towards the discretization procedure. The test is totally void of issues of model quality. An in-sample stable model can be arbitrarily bad. Out-of-sample stability is also mainly an issue concerning the model itself, but has aspects of something more in it.

Out-of-sample stability means that if you calculate the true objective function value corresponding to the solutions coming from different scenario trees, you get (about) the same value. In other words,

$$f(\hat{x}_i; \boldsymbol{\xi}) \approx f(\hat{x}_j; \boldsymbol{\xi}).$$

Hence, it is not such that your scenario generation procedure has generated a stability that is not really there. This could, for example, happen if your scenario generation procedure consistently avoided a difficult tail of the distribution, so that the in-sample stability you had observed was really just over a part of the support of the random variables. And the in-sample stable solutions you had found behaved very differently on the part of the support you had avoided. The true value showed you this, and you ended up out-of-sample unstable.

The starting point was that the true problem "$\min_x f(x; \boldsymbol{\xi})$" could not be solved. What we need to test in out-of-sample stability is slightly different, namely to fix x at $\hat{x}_i$ but then take expectations with respect to the true distribution. This is in many cases doable. If $\boldsymbol{\xi}$ is discrete (but with far too many scenarios to solve the corresponding true optimization problem) you may still be able to calculate $f(\hat{x}_i; \boldsymbol{\xi})$ as that just amounts to solving a large number of second-stage problems. If $\boldsymbol{\xi}$ is continuous (or discrete but too large), the obvious method is to sample from the distribution in order to approximate the true value.

In many cases, the correct way to calculate the out-of-sample value is to construct a simulation model of your problem. That way, the out-of-sample calculations will not only take care of the fact that $\mathcal{T}$ is an approximation, but also the fact that f most likely only approximates the true problem as well. In a simulation model, you can normally add more details than in an optimization model.

You would also wish to have the objective function values coming from in- and out-of-sample stable solutions being the same. Otherwise, there is something going on you have not understood. Why are they different?

If we cannot evaluate $f(\hat{x}_i; \boldsymbol{\xi})$, there is another, weaker, out of-sample test we can perform: let us have two scenario trees $\mathcal{T}_i$, $\mathcal{T}_j$ with corresponding optimal solutions $\hat{x}_i$, $\hat{x}_j$. If the model is out-of-sample stable, we should have

$$f(\hat{x}_i; \mathcal{T}_j) \approx f(\hat{x}_j; \mathcal{T}_i).$$

4.2.3 Bias

Even if you have both in- and out-of sample stability, you may be in trouble. And this trouble is *bias*. You may be stable, but bad. This is not easy to test for in a simple way. You need to check if

$$\min_x f(x; \boldsymbol{\xi}) \approx \min_x f(x; \mathcal{T}_i), \quad \text{for all } i$$

Only if you can solve the true problem can you make this test. Occasionally, your true distribution may be discrete, have very many scenarios, but even so, maybe you could solve the model once (using unreasonably much time) just to check.

Of course, if you have two scenario generation procedures for a given algebraic model, and both are in- and out-of-sample stable, but the true objective function values are different, then at least one of the generation procedures is biased.

To try to answer questions about quality other than stability you will need to turn to statistical methods that we shall discuss a bit later.

4.2.4 Example: A Network Design Problem

Chapter 5 discusses a network design problem. In the model's early development we ran into the problem of in-sample instability. This showed that something was wrong with the model (meaning the algebraic model plus the scenario generation procedure). The problem turned out to be that arbitrarily small changes in maximal demand could result in extra capacity being bought in the network. As capacity came in large pieces (trucks, in fact), these extremely small changes in demand resulted in big jumps in costs. Hence, no stability. Therefore, something had to be done. An in-sample unstable model is useless. What we did was to extend the scenario generation model to include the requirement that outcomes had to be within the defined support. Stability was achieved. This does not imply that we now had a good model. But at least the model now could be further tested. An in-sample unstable model cannot be properly tested, as it produces random results. For more details, see Chap. 5.

4.2.5 The Relationship Between In- and Out-of-Sample Stability

Despite the discussion above, there is not principally an ordering of stability results as we have indicated. It is, in fact, possible, at least in principle, to have out-of-sample stability without in-sample stability. That would mean that despite the fact that the different scenario trees imply rather different solutions and optimal function values, the true, out-of-sample, objective function values are (almost) the same. If that is the case, it would be possible to proceed with the model, but in a very careful manner. A solution corresponding to an arbitrary tree would then be as good a solution as that based on any other tree, but the corresponding optimal function value would not be a good measure of the true value. It should be clear that many types of testing of the model would be very difficult, though possibly not impossible, in such a setting. We would, however, not in any way recommend to proceed with a model that does not demonstrate in-sample stability.

> If you do not have in-sample stability, you have not properly understood your model.

4.2.6 Out-of-Sample Stability for Multiperiod Trees

So far, we have only discussed the single-period case. For multiperiod trees, the situation is more complicated: we cannot simply take a solution based on one tree and evaluate it on another one, as the nodes beyond the root do not coincide. The same is true for a simulation model: the simulated scenarios will not coincide with the tree beyond the root, so we will not be able to use the found optimal values except in the root.

This, however, is exactly what happens in reality as well: only the first-period "root node" solution gets implemented. When we need another decision later, we re-run the model with an updated tree and implement the new root decision. This could be simulated using a rolling-horizon approach, see [31] for details.

If we do not have a simulator, we can at least do the following: build two trees $\mathcal{T}_1$, $\mathcal{T}_2$ and find the corresponding solutions $\hat{x}_1$, $\hat{x}_2$. Then solve the optimization model on tree $\mathcal{T}_1$ with the root values fixed to the root values of $\hat{x}_2$, and vice versa. If the method is out-of-sample stable, we should get approximately the same optimal objective values.

4.2.7 Other Approaches to Stability

There is also another approach to measuring stability of stochastic programs, by estimating the error caused by using distributions that only approximate the true one. Interested readers can have a look at, for example, Dupačová and Römisch [9] and Fiedler and Römisch [14] for two-stage problems, and Heitsch et al. [21] for multistage problems.

4.3 Statistical Approaches to Solution Quality

If you have a solution, wherever it comes from, you may want to ask how good it is relative to the unattainable optimal solution. Out-of-sample calculations can estimate the optimal value associated with a solution, and if you have several candidates, you can use out-of-sample calculations to pick the best one. You may have established in- and out-of-sample stability as we already discussed, but as there is a potential problem of bias you may be worried. If stability has not been established the question is even more pressing: Is this actually a good solution?

Please note that this section is technically more challenging than most others in this book.

4.3.1 Testing the Quality of a Solution

So far, we have been concentrating on *stability* and avoided discussions about solution quality. The reason is that without stability, quality is not well defined as it is different every time we run the scenario-generation and optimization models. The other reason is that estimating quality of a given solution $\tilde{x}$ is difficult, as it amounts to evaluating the *optimality gap* (approximation error)

$$\mathrm{err}(\tilde{x}) = f(\tilde{x}; \boldsymbol{\xi}) - \min_x f(x; \boldsymbol{\xi}). \tag{4.3}$$

Since the right expression is the reason why we need scenarios in the first place, it might seem impossible to evaluate the above expression. There is, however, a way to derive a *statistical estimate* of the error, as long as we can sample from the underlying distribution $\boldsymbol{\xi}$, and the objective function f is an expected value, i.e.,

$$f(x;\boldsymbol{\xi}) = E^{\boldsymbol{\xi}}[F(x,\boldsymbol{\xi})] \qquad \text{and} \qquad f(x;\mathcal{T}) = \frac{1}{N}\sum_{s\in\mathcal{S}} F(x,\mathcal{T}^s), \qquad (4.4)$$

where N is the number of samples (scenarios) in $\mathcal{T}$ and $\mathcal{T}^s$ is the s'th scenario in $\mathcal{T}$.

This approach comes from Mak et al. [43]: we start with a sampled tree $\mathcal{T}_i$ with S scenarios. From (4.4) is follows that $f(x;\mathcal{T})$ is an unbiased estimator of $f(x;\boldsymbol{\xi})$, i.e.,

$$E[f(x;\mathcal{T})] = E\left[\frac{1}{N}\sum_{s\in\mathcal{S}} F(x,\mathcal{T}^s)\right] = f(x;\boldsymbol{\xi}). \qquad (4.5)$$

In addition, one can show that

$$E\left[\min_x f(x;\mathcal{T})\right] \leq \min_x f(x;\boldsymbol{\xi}), \qquad (4.6)$$

that is the scenario-based solutions are, on average, optimistic. Combining the last two equations with (4.3) gives the following:

$$\begin{aligned} \text{err}(\tilde{x}) &= f(\tilde{x};\boldsymbol{\xi}) - \min_x f(x;\boldsymbol{\xi}) \\ &\leq E[f(\tilde{x};\mathcal{T})] - E\left[\min_x f(x;\mathcal{T})\right] \\ &= E\left[f(\tilde{x};\mathcal{T}) - \min_x f(x;\mathcal{T})\right], \end{aligned}$$

which can be also written as

$$\text{err}(\tilde{x}) \lessapprox f(\tilde{x};\mathcal{T}) - \min_x f(x;\mathcal{T}). \qquad (4.7)$$

This means that we have found a *stochastic upper bound* for the approximation error. Of course, having just one estimate of the sample error is not enough; what we need is to repeat the whole procedure for M such trees $\mathcal{T}_i$, which would give us a better estimate

$$\text{err}(\tilde{x}) \lessapprox \frac{1}{M}\sum_{m=1}^{M}\left[f(\tilde{x};\mathcal{T}_m) - \min_x f(x;\mathcal{T}_m)\right], \qquad (4.8)$$

where we can replace "$\lessapprox$" by "$\leq$" if we let $M \to \infty$.

In addition to the point estimate, it is also possible to use the n values to derive a confidence interval for $\text{err}(\tilde{x})$, which could also be improved by various variance-reduction techniques—we refer interested readers to [43]. There are also variants of the estimate that require only two or even only one replication, see Bayraksan and Morton [3].

To get a better understanding of the quality of the estimator and how it can be improved, we can rewrite (4.8) as

$$
\begin{aligned}
\text{err}(\tilde{x}) \lessgtr\ & \frac{1}{M} \sum_{m=1}^{M} f(\tilde{x}; \mathcal{T}_m) - f(\tilde{x}; \boldsymbol{\xi}) \\
& + f(\tilde{x}; \boldsymbol{\xi}) - \min_{x} f(x; \boldsymbol{\xi}) \\
& + \min_{x} f(x; \boldsymbol{\xi}) - \frac{1}{M} \sum_{m=1}^{M} \min_{x} f(x; \mathcal{T}_m).
\end{aligned}
\tag{4.9}
$$

Since the second element in the right-hand side is equal to $\text{err}(\tilde{x})$, this gives the error of the estimator (4.8) as

$$
\begin{aligned}
\text{est. error} =\ & \frac{1}{M} \sum_{m=1}^{M} f(\tilde{x}; \mathcal{T}_m) - f(\tilde{x}; \boldsymbol{\xi}) \\
& + \min_{x} f(x; \boldsymbol{\xi}) - \frac{1}{M} \sum_{m=1}^{M} \min_{x} f(x; \mathcal{T}_m)
\end{aligned}
\tag{4.10}
$$

The first element has zero expected value from (4.5), with variance decreasing with increasing M and the number of samples in $\mathcal{T}_m$. The second element is the *bias* of the estimator—it is positive because of (4.6) and can be decreased only by improving the solutions of "$\min_{x} f(x; \mathcal{T}_m)$", i.e., by increasing the size of $\mathcal{T}_m$.

Another important observation is that the first element of the gap estimator (4.8), is equivalent to evaluating the solution $\tilde{x}$ using $\mathcal{T}_{N'}$ with $N' = M \times N$ samples. From (4.9), we then see that this is used as an estimate of the true objective value $f(\tilde{x}; \boldsymbol{\xi})$; it follows that we could use any tree with $N' \gg N$ samples, i.e., use

$$
\text{err}(\tilde{x}) \lessgtr f(\tilde{x}; \mathcal{T}_{N'}) - \frac{1}{M} \sum_{m=1}^{M} \min_{x} f(x; \mathcal{T}_m)
\tag{4.11}
$$

instead. There is, however, a small catch here: using the same samples in both elements of (4.8) results in *variance reduction*, so we would probably need to use $N' \gg M \times N$ to get results with the same variance as in (4.8).

4.3.2 Solution Techniques Based on the Optimality Gap Estimators

So far, we have used the gap estimators only to evaluate the quality of a given solution $\tilde{x}$. However, since they allow us to estimate not only how good a given solution is, but also how far from the optimal solution it is, it should not come as any surprise that these estimates form the basis for several methods for *solving* stochastic-programming problems.

The most well-known of these is the so-called SAA method from Kleywegt et al. [38] and it works as follows: we sample M trees $\mathcal{T}_m$ with N scenarios and for each of them estimate the optimality gap (4.8) or (4.11) with some chosen N'. If both are sufficiently small, we pick the best one as our solution; otherwise, we increase M, N, or N' and repeat the whole procedure. There are, of course, many important details such as the choice of M, N and N', and the exact meaning of "sufficiently small", for which we refer interested readers to the original paper.

Another such method can be found in Bayraksan and Morton [4]. This method builds on the assumption that we can get a sequence of feasible solutions x_k that converges to the optimal solution of (4.2)—such as solutions of the scenario approximations (4.1) with increasing sample sizes. The method itself is then similar to the SAA method with M equal to one: for a given k, we test whether the solution x_k is "good enough" using one tree with N_k scenarios. If yes, we stop; if not, we generate the next solution x_{k+1} and repeat the procedure. Note that since this method estimates gaps using only one tree at each step, it is based on gap estimators from [3] instead of (4.8).

It may appear that these methods always work. But that is not the case, and should in fact not surprise you: your problem may be so large or complicated that there is no way with today's technology we can find good solutions. The methods assume that there is a reasonable chance that a sampled tree leads to a good solution. But for a really large problem that probability might be practically zero (as you really need scenario trees vastly larger than what you can handle), so you just keep wandering around in darkness.

Sample based methods using statistical approaches to check solution quality may not work if the problem is too large.

An interesting observation about the above method is that it is not important where the solutions come from—we only need to guarantee that we have a sequence converging to the optimal solution. So if we have a scenario-generation method that is better than sampling that converges to the true distribution, it is probably a good idea to use it instead. But what if we have a scenario-generation method that is, according to the tests presented in this chapter, better than sampling for the range of tree sizes we are able to solve, but we do not know if it converges to the true distribution (or even know it does not)? Can we still use it to generate the solutions, instead of using sampling? We are not aware of any studies on this subject, but we dare guess that the answer is "yes"—scenario-tree quality for the tree sizes we actually use should be more important than the limit behaviour we never observe.

4.3.3 Relation to the Stability Tests

At first sight, this statistical approach does not seem to be in any way related to the stability we talked about earlier in this chapter. This, however, is not correct: there is a very close relationship between these two, as we will show below.

The statistical tests help us to evaluate the *bias* from Sect. 4.2. There, we first require a given scenario-generation model to be *stable*, because otherwise we cannot say what a "solution" or "optimal objective function value" is, as they are different on every run of the model. Then we explain that stability is not sufficient to guarantee good solutions, since a wrongly selected scenario-generation method can still lead to sub-optimal solutions of the optimization problem—in which case we called the scenario-generation method "biased". However, back in Sect. 4.2.3, we did not have any tools to estimate the bias. The statistical tests from Sect. 4.3 allow us to estimate at least an upper bound on the optimality gap (bias) and in this way complement nicely the other tests from Sect. 4.2.

The statistical tests provide also lend support to the intuitive idea that having a big difference between the in- and out-of-sample objective values $f(\hat{x}; \mathcal{T})$ and $f(\hat{x}; \boldsymbol{\xi})$, for a solution $\hat{x} = \min_x f(x; \mathcal{T})$, is not a good sign: from (4.7), we see that the difference is a stochastic upper bound on the optimality gap,

$$f(\hat{x}; \boldsymbol{\xi}) - f(\hat{x}; \mathcal{T}) = f(\hat{x}; \boldsymbol{\xi}) - \min_x f(x; \mathcal{T}) \gtrless \mathrm{err}(\hat{x}).$$

We should, of course, keep in mind that since it is a upper bound, a big difference does not necessarily mean big optimality gap; as we can see from (4.9) with $M = 1$, the difference can also be caused by the bias of the gap estimator.

In addition, we can see what the stability concepts tell us about the gap estimator itself. In this context, the *in-sample stability* from Sect. 4.2.1 refers to the variance of "$\min_x f(x; \mathcal{T}_m)$" from (4.8); if it is big (i.e., if we do not have stability), then the whole estimator (4.8) will have a big variance and hence a confidence interval that is too wide for the estimate to be useful. Similarly, the *out-of-sample stability* from Sect. 4.2.2 refers to the variance of $f(\hat{x}_m; \boldsymbol{\xi})$, approximated by $f(\hat{x}_m; \mathcal{T}_{N'})$—which is the first element in the optimality-gap estimator (4.11). Again, it follows that out-of-sample instability implies big variance of the gap estimator, with the same results as in the in-sample case. Finally, the *bias* from Sect. 4.2.3 does not exist in this context, since the estimators use sampling, which is unbiased—see (4.5).

4.4 Property Matching Methods

We mentioned earlier that we should measure the quality of a scenario tree using the objective function of the stochastic program as metric. That is, if two discretizations give the same expected cost (or profit) they are equally good, and the optimal discretization is one that trades off problem size (the number of scenarios) and quality of solution as well as possible. So for a given number of scenarios we want as good an approximation of expected costs as possible, or, for a given quality of the expected cost, we want as few scenarios as possible. This is reflected in our quality tests in the previous section, they all use the objective function to measure quality.

In Sect. 4.2 we showed how to test the quality of a scenario generation method by studying stability. The idea was to establish a situation where you could safely run your scenario generation method only once and then solve the stochastic programming problem, being confident that the solution you got made sense. This scenario generation method could be sampling, but we indicated many times that for a reasonably large problem sampling is not likely to lead to stability. Rather, for sampling it is more reasonable that you need results from Sect. 4.3 to check the quality of the solutions you obtain.

If you have your own scenario generation method, you are advised to use the stability tests first to see if you can avoid the heavy work of testing the quality of solutions you obtain. If you cannot establish stability, then testing quality is your only option if you at all care about the meaning of what you have obtained.

In this section we shall present an approach for generating scenarios which is an alternative to pure sampling, and which tries to take into account that the quality of a scenario tree should be measured by the stochastic program you are solving, not by purely comparing distributions. Although the tests in Sects. 4.2 and 4.3 use the optimization problem to measure stability and quality, if you used sampling to generate the trees, you did not use a scenario generation procedure that tried to fit the test; when you sample you indirectly care about all aspects of a distribution, also those that do not matter to the optimization problem and hence the tests. This is no problem if you end up with positive results. But it is rather likely that this lack of match between how you generate and how you measure scenarios is a cause for trouble.

But is it better to match moments (and other statistical properties) than to sample and in fact try to match *all* properties? The answer is: only if you end up regarding those properties that matter. How can we know that apart from special cases like the Markowitz model we illustrated in Fig. 4.1? Rarely can we do it based on such direct analysis. But we can use our brains a bit:

- The means and variances are almost always important
- If we use a risk measure that is not symmetric (contrary to the variance which is symmetric), the third moments are very likely important.
- If we care about the probability of really bad outcomes, we may have to look at the fourth moments which describe the thickness of the tails.
- We need to ask if the distributions are elliptic (like the normal) so that correlations are likely to describe co-variation well. If the answer is no, we need to consider whether or not that matters.

In what follows we shall first discuss how you might think about formulation a model that produces a scenario tree matching a variety of possible properties (not just moments). Numerically that approach is not very good in light of stability, but we show it as it illustrates the idea behind property (moment) matching. Then we pass to an approach that looks at only four marginal moments plus correlations. That method is efficient and simple to use. It is our experience that it very often leads to stability—but that must of course be tested.

If you do not get stability with this approach, there are two possibilities: make the tree larger or revise your choice of properties. There are no automatic ways of doing the latter, you need to use problem understanding and simply try. There is of course a reasonable chance that also for this kind of approach, just as for pure sampling, you need so many scenarios to get anything meaningful, that the problem cannot be solved. Do not be surprised or discouraged by this. Some problems are simply so large that it isn't possible with today's technology to solve meaningful stochastic versions of them. In this case you will either have to develop stronger algorithms for solving stochastic programs, probably specialized for your problem (which certainly is a feasible path in many cases), or you will have to revise and simplify your model.

4.4.1 Regression Models

One way to set up a scenario tree is to define properties you want to be present in the tree, and then, for example, by non-linear programming, create a tree with those properties. In such a problem, the variables are the outcomes of the random variables and possibly also the probabilities. The ideas here are taken from [27].

Let $\boldsymbol{\xi}$ be the discrete random vector we are to generate. A single random variable is then denoted $\boldsymbol{\xi}_i$, for $i \in \mathcal{I}$. A scenario is a vector ξ with individual elements ξ_i for $i \in \mathcal{I}$. If we need to index a scenario, it will show up as ξ^s with elements ξ_i^s for $s \in \mathcal{S}$. Assume you want a specific mean μ in your scenario tree. Hence, you want

$$\sum_{s \in \mathcal{S}} p^s \xi^s = \mu.$$

You may think of this as either a constraint in a non-linear optimization problem (with ξ and p as variables), or you may put the expression into the objective function with a penalty for violation.

Variances could be controlled by

$$\sum_{s \in \mathcal{S}} p^s (\xi_i^s)^2 - \left(\sum_s p^s \xi_i^s\right)^2 = \sigma_i^2.$$

In the same way, you may proceed with any expression that can be written as a function of ξ and p. For example, if you need to ensure that in at least one scenario, $\boldsymbol{\xi}_i = \underline{\xi}_i$, the relevant expression is

$$\xi_i^1 = \underline{\xi}_i.$$

As the indexing of the scenarios is arbitrary, we have simply chosen to let the extreme value of ξ_i be in scenario 1. Any choice would be OK. As a final example, if you wish the correlation between ξ_i and ξ_j to be r_{ij}, you need

$$\sum_{s \in S} p^s \, \xi_i^s \xi_j^s - \mu_i \mu_j = \sigma_i \sigma_j \, r_{ij} \,,$$

as long as also means and variances are controlled.

If the probabilities are variables, you also need to add $p^s \geq 0$ and $\sum p^s = 1$. Numerics may dictate that it is simpler to let the probabilities be fixed in the scenarios, typically at $p^s = 1/|S|$. The price to be paid for this simplification is that you may need more scenarios to describe what you want.

Whatever choices you have made, you now need to find a set of p^s and ξ^s that is feasible (with respect to whichever expressions you have chosen to add as constraints) and with zero as the optimal objective value. If that is achieved, you have a scenario tree with exactly the required properties.

But before embarking on a regression-based approach, see if you can utilize the special situation described in Sect. 4.4.2. The reason for that is twofold. Firstly, it is much simpler numerically. Secondly, the approach from Sect. 4.4.2 produces better results. The reasons for this are discussed under "A different interpretation of the model" on page 99. However, regression models form the basis for all these ideas, and there might be cases where that is your only choice, as you need to represent distributional properties for which there are no simple approaches.

4.4.2 The Transformation Model

This is an iterative procedure that produces trees with specified first four marginal moments and correlations outlined in [28]. It is based on transformations, and is very quick. Although there has been some work on extending the method beyond the four moments and correlations, there is clearly a limit to how far it can be extended.

The procedure consists of three parts: an initialization, and two transformations that are repeated in an iterative loop.

4.4.2.1 Initialization

The procedure needs a starting distribution. Although, in principle, you can start many different ways, we usually start with independent standard normal random variables ξ_i.[1] While it would be easy to obtain the values for

---

[1] We do not recommend using the starting distribution described in [28], as it seems to have a negative effect on the quality of resulting scenarios—not to mention that it does not have to exist in the first place.

each margin by sampling, we prefer to use a fixed discretization to increase the "smoothness" of the distribution. For discretization with $|\mathcal{S}|$ scenarios, we would typically divide the support into $|\mathcal{S}|$ intervals, each having the same probability, namely $1/|\mathcal{S}|$. The outcome associated with an interval is the conditional expected value. Then, create scenarios $\xi^s = (\xi_1^s, \xi_2^s, \ldots, \xi_{|\mathcal{I}|}^s)$ by randomly combining outcomes from the different discretizations. Here ξ_j^s is one of the outcomes for random variable j. Since the combinations are random, all correlations in the tree will have zero expectations. However, since the tree is of limited size, the actual correlations will be somewhat off zero.

So, with this approach, the starting tree has approximately uncorrelated standard normal margins. Note that if you have other ways of initiating the distributions, for example, ways of better representing co-variation among the random variables, you should test them.

4.4.2.2 Correcting Correlations

The first transformation changes a tree to obtain the desired correlations. To do it, we use the standard method for generating correlated normal variates: if $\boldsymbol{x}$ is a vector of uncorrelated random variables, and R is a correlation matrix with Cholesky decomposition $R = LL^T$, then $\boldsymbol{y} = L\boldsymbol{x}$ has a correlation matrix equal to R.

If $\boldsymbol{x}$ was a vector of standard normal variables, the margins of $\boldsymbol{y}$ would again be standard normal. In the general case, however, the transformation changes the marginal distributions (and hence the moments). The only thing we can guarantee is that if all the margins of $\boldsymbol{x}$ have zero mean and variance equal to one, the same will be true for the margins of $\boldsymbol{y}$. To take advantage of this, all the random variables are standardized (set to zero mean and variance equal to one) inside the algorithm. They are transformed to their specified means and variances at the very end of the algorithm, using a simple linear transformation $\boldsymbol{y}_i \leftarrow \mu + \sigma\,\boldsymbol{y}_i$.

The transformation $\boldsymbol{y} = L\boldsymbol{x}$ works only if the margins of $\boldsymbol{x}$ are uncorrelated. In the first iteration, they may be close, in the sense that we expect all correlations to be zero. In later iterations, that is certainly not the case. Assume we want $\boldsymbol{y}$ with correlation matrix $R = LL^T$, but have $\boldsymbol{x}$ with correlation matrix $R_k = L_k L_k^T$. Then we know that $\boldsymbol{z} = L_k^{-1}\boldsymbol{x}$ is uncorrelated, and

$$\boldsymbol{y} = L\boldsymbol{z} = (LL_k^{-1})\boldsymbol{x}$$

has the required correlation matrix R. Hence, it is the matrix LL_k^{-1} which is used in the computations. The index k here plays the role of an iteration counter.

Note that the Cholesky components L and L_k have a "1" in the top left corner, so the same must be true for the lower-triangular matrices L_k^{-1} and LL_k^{-1}. This means that the first margin is never changed by the transformation.

4.4.2.3 Correcting Moments

This is the most computing intensive part of the algorithm. A cubic transformation of the outcomes in the tree, changes the tree into another tree with the required first four marginal moments. The setup is that we, for each random variable i, need to find four parameters a_i, b_i, c_i, d_i such that

$$\boldsymbol{y}_i = a_i + b_i \boldsymbol{x}_i + c_i \boldsymbol{x}_i^2 + d_i \boldsymbol{x}_i^3$$

has the specified first four moments. The parameters are found by solving a system of four implicit equations in four unknowns. For details on the equations we refer to [28].

Since we use a non-linear transformation it will change the correlations, leaving us with margins with the correct moments and a slightly-off correlation matrix—so we use the correlation-correcting transformation again. This is repeated a few times until both moments and correlations are in place.

4.4.2.4 Illustrative Example

As an example of the algorithm, we generate 200 scenarios for two random variables with correlation 0.5. The first four moments of the first one are {0,-1,0.75,4}, the other {0,1,0.75,4}.[2] Since the first variable does not change by the correction the of correlations (see above), it remains unchanged once its moments are corrected in the first iteration. We thus present only a histogram of the second variable, together with a scatter-plot to show the two-dimensional structure.

The results of the algorithm are in Fig. 4.2. The first line shows the distribution after the first correction of the correlation, so the margins are still normal. The second line presents the situation at the end of the first iteration, after the correction of the marginal distributions. Note that the correlation is slightly distorted. The last two lines show the corresponding results of the second iteration. We can see that correcting the correlation slightly distorted the margins, but the following correction of the moments changed the correlation so slightly that it remained correct up to the second decimal place.

With only 200 scenarios it is impossible to present the shape of the density, so we have done the same test with 10,000 scenarios. The algorithm was again stopped after the second iteration, and the density was estimated using a kernel estimator. The result is in Fig. 4.3.

4.4.2.5 Does It Always Work?

There is no proof that the above iterative procedure works in all cases. However, there is no reason to be very concerned. It is very easy to check after

---

[2] For moments of order $k > 2$, we use $\mathrm{mom}_k(\mathbf{X}) = \boldsymbol{E}\left[(\mathbf{X} - \mu)^k\right]/\sigma^k$.

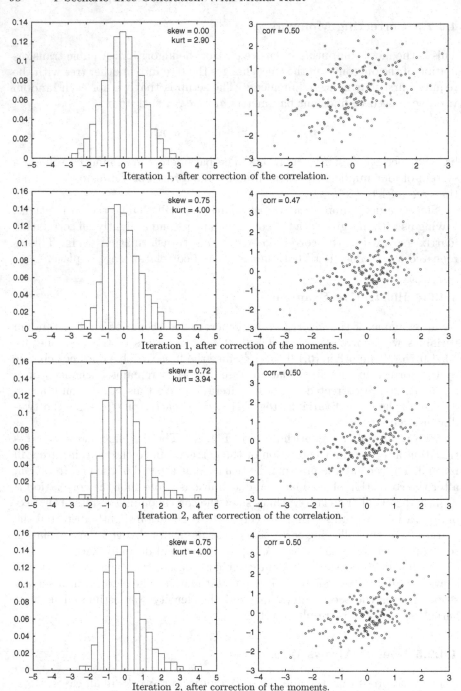

Fig. 4.2: First two iterations in a case of two correlated variables

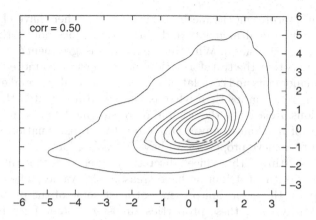

Fig. 4.3: Iteration 2, after correction of the moments—10,000 scenarios

the procedure has stopped what are the statistical properties of the resulting tree. Hence, there is no danger that we end up using a tree with incorrect properties.

But is it possible that the transformations will yield incorrect results? So far, we have never observed lack of convergence, except in two situations. The first is that the properties used to define the tree are internally inconsistent, that is, that there exists no tree like the one asked for. It is possible to define values for the marginal moments which do not correspond to anything real. This will of course never happen if the numbers are calculated from a single actual data set. But if they are simply defined, that could happen. The second case which might result in an incorrect tree is that the tree we ask for is too small. Clearly, the more properties you ask for, then more variables (i.e., scenarios) you need.

Hence, if you observe lack of convergence, the obvious first attempt around it would be to make the tree larger, and try again. If that does not work, it is worth checking if the required properties make sense at all. Although never observed, there is of course (as long as we have no proof) a chance that lack of convergence may show up even with a well defined distribution and large enough tree.

In the end, the only way to find out how the method works for *your* problem is to test it yourself. It is freely available from http://work.michalkaut.net/#HoylandEA03, including the source code.

4.4.2.6 A Different Interpretation of the Model

The interpretation given so far of the transformation model is simply that it is an efficient way of solving the regression model for a specific case, namely four marginal moments and correlations. However, that may not be the most useful way of interpreting the method. Several researchers have claimed that matching four marginal moments and correlations does not work (meaning does not give useful results). Why this apparent disagreement?

The reason is that the transformation model does much more than match four marginal moments and correlations. First of all, it allows (and encourages) to generate more scenarios than what is needed to just match the moments and correlations. Relative to the original regression model (which does not use extra degrees of freedom in any useful way), the transformation model uses this freedom to achieve properties we have not asked for.

For once, starting with a "nice" discretization ensures that many unspecified properties of the distribution have "reasonable" values. The transformations involved in the algorithm preserve at least some of the properties and, more importantly, leave these properties almost the same from iteration to iteration of the scenario generation procedure.

The resulting marginal distributions are "smooth"—as a visual statement of how they look. Hence, the resulting scenario tree can also be seen as an approximation of the whole distribution, with some properties guaranteed and some at least having "reasonable" values. And it is this total package, rather than just the limited set of properties, that produces stable models.

You may be disturbed by the fact that we do not quite know what is going on. It would certainly be better if we knew, but one of the purposes of Sect. 4.2 is to test (rather than simply discuss) if a certain procedure produces stable results. And it turns out that for many models, the transformation model produces stable results, while other methods that match four marginal moments and correlations do not.

4.4.2.7 Extensions of the Model

While the method works satisfactorily for many models, there are situations where four moments and correlations simply do not provide sufficient control over the distribution. As an example, the model might be sensitive to the shape of the multivariate distribution, in which case correlations would not provide enough control. Recently, there has been some development to cope with both possibilities.

In the case when we need more than just correlations to control the co-variation, [32] discusses controlling the shape of the multivariate distribution using the notion of copulas. Unlike the previous case, which basically generalizes the transformation method, here the approach is rather different. In addition, the proposed methods do not allow for an exact control of correlations, though this can be solved by using the transformation method as a post-process.

4.4.3 Independent and Uncorrelated Random Variables

In some cases the random variables are genuinely independent. However, more commonly we get independent or uncorrelated random variables when we apply such algorithms as principle component analysis to the underlying randomness. In some respects independence is easier to handle, but most of the time it is not.

The case where independence is easy is as follows. Assume you have n random variables, each taking on k possible values. If we make scenarios by combining all-against-all, arriving at k^n scenarios, it is very easy to calculate scenario probabilities in the independent case.

But we can rarely allow ourselves the luxury of having k^n scenarios. Hence, we must resort to ideas like those presented in this chapter. And then independence is a property we *enforce* on the scenarios (in fact we are only able to enforce zero correlations). Since the method, as explained in this chapter starts out with almost zero correlations, it will converge quickly in that case. But otherwise, independence (or zero correlations) is not a major issue, any correlation structure can be enforced by this method.

4.4.4 Other Construction Approaches

In his famous paper [45], Pflug showed that the approximation error can be bounded by the distance between the discretization and the original distribution, expressed using the Wasserstein metric (transportation distance), and proposed to generate scenarios by constructing discretizations that minimize this distance. This gave rise to a series of papers about generating or improving scenario trees by minimizing this—or some related—metric; the most well-known are the papers about *scenario reduction* methods, see Dupačová et al. [10], Heitsch and Römisch [20, 22] for two-stage, and Heitsch and Römisch [23] for multistage models. These methods give much better results than pure sampling, for a given number of scenarios.

Note however that these methods focus purely on some distance between the two distributions; the optimization problem appears in the definitions of the approximation error, but disappears in the bounding process. This means that these methods will always try to replicate the whole distribution, instead of focusing only on the properties important for the given optimization model (like mean and (co)variances for the Markowitz model)—and that does not come for free in terms of the number of scenarios. On the other hand, removing the optimization model from the scenario-generation process can be seen an advantage, as it means that scenario generation can become a specialty in its own right, not requiring optimization knowledge. Many prefer these approaches for that reason.

In the end, the decision is the same as for so many other methods: we should test if they work for our problem, using either the stability or statistical tests, and if they do then all is fine.

4.5 Choosing an Approach

So how should you go about this? If you are going to solve a certain problem only once, trying SAA first is probably a good choice. You may have to solve your model quite a number of times, but most likely that is the easiest way forward. But by now you understand that this might lead nowhere if the number of scenarios you can handle numerically is too low for the problem you solve to have any reasonable relationship to what you want to solve. This is the normal difficulty of sampling.

But if the model is to be solved repeatedly, the conclusion is not so obvious even if SAA generates reasonable results. If this is a model of the type that is solved every morning, say a production model, you are probably rather constrained on elapsed time. That may in its own right rule out SAA if you at the same time really care about the quality of the solution. If you instead use, for example, moment matching and establish in- and out-of-sample stability (and maybe even use one of the statistical approaches to get a feeling for statistical quality, but do it off-line) you can then proceed day-by-day by having to solve just one stochastic program. For sure this is a heuristic, but not one without any foundation in reality.

If you are in an academic setting, and need to solve hundreds or thousands of similar problems to prove, for example, some properties of the solutions or solution procedures (as is done in the thesis underlying Chap. 5 as well as many, many other pieces of academic work), you would also hesitate on using SAA as each such case would then require a repeated solve of your stochastic program to establish statistical convergence. The elapsed time would easily be enormous. We are then assuming, of course, that you care that the problems you solve make any sense. So instead we would suggest that you spend a lot of time up front on figuring out how to generate scenarios so as to obtain stability, thereby reducing the solution times of the repeated cases substantially.

Chapter 5

Service Network Design: With Arnt-Gunnar Lium and Teodor Gabriel Crainic

> The shortest distance between two points is under construction.
> – *Noelie Altito*

This chapter represents an investigation following the lines of this book, where the focus is that of a graduate student studying the effects of uncertainty on a specific problem. There is no customer in this problem, and it has not reached the level of sophistication needed for a real application. However, it goes to the heart of this book: What does stochastics do to my problem? What are the implicit options? This chapter is based on the Ph.D. thesis of Arnt-Gunnar Lium of Molde University College. For an overview see [40]. You are going to meet an inherently two-stage problem with, principally, infinitely many stages. However, since in this situation we do not really need the decisions of the inherent second stage, we can approximate, ending up with a two-stage model. In our view, this points to the heart of stochastic programming: inherently two-stage problems with rather complicated stages after the first one.

5.1 Cost Structure

The starting point here is the problem of less-than-truckload trucking (LTL), that is, trucking services set up to service the market of small packages, where renting a full truck is out of the question. The problem facing the planner is to decide how many trucks to buy (or rent) and which schedules to set up for these trucks. In the example we present here, the trucks have weekly schedules. As flows of goods are usually quite unbalanced, the movement of empty trucks for repositioning is crucial to capture. Also, trucks may at times be parked during full days, simply because they are not needed that day. We will let the daily rate of having a car parked be lower than that of having it run.

This cost structure has two different interpretations. We can either assume we rent the trucks (long term, not day-by-day), and the cost of having the truck parked equals the pure renting costs. Then, for each of the movement

A.J. King and S.W. Wallace, *Modeling with Stochastic Programming*,
Springer Series in ORFE, DOI 10.1007/978-0-387-87817-1_5,
© Springer Science+Business Media New York 2012

costs, by subtracting the cost of a parked truck, we find the variable costs of actually running the truck along a certain route. Alternatively (since we work by the week), seven times the daily parked rate can be interpreted as the weekly fixed costs of having a truck, and the differences, just as above, are the variable costs. Hence, the model covers both a pure variable cost interpretation and a fixed- and variable-cost interpretation. If you only need to pay for trucks that are actually being used, then the problem changes considerably, and that version is not covered in this chapter.

5.2 Warehouses and Consolidation

A package does not necessarily follow a given truck from its origin to its destination. Rather, the whole point of the service network design is to set up truck schedules so as to facilitate transport by using warehouses and consolidation of goods. So packages may spend some of their time in one truck, arrive at a warehouse, maybe stay some time at a warehouse, and then continue with another truck. It can, in principle, spend time on many trucks before arriving. Each package has, in addition to a starting and ending node, a day of being available in its starting node and a latest day of arrival. Clearly, the longer you are allowed to keep the package, the more "interesting" routes you can find to efficiently use your trucks' capacity and, therefore, the more money you can make. On the other hand, in reality, the quicker you get the package to its destination, the higher the prices you can charge. In this model, a package simply has a date of arrival at its starting node and a due date. But while modeling, you must consider the foregoing trade-off.

5.3 Demand and Rejections

The next major modeling issue here is that of how to model orders and potential rejections. This is, from a modeling perspective, rather complicated, and we would clearly advise you to think carefully about it. You may have a history of demand. Should that be used as a measure of potential future demand? If you answer yes, you may be indicating that even though you are about to create a new (and better?) service network (new schedules), you will observe the same demand as before. Is that reasonable? You may, of course, also look at the full potential demand for your services, also including the demand presently being picked up by competitors. But is that reasonable? Probably not.

Let this question of how to model the demand rest a short while, and let us look at another aspect of the model—as they are connected. Will you reject orders that are not profitable or that you have no capacity to carry? Traditionally, deterministic LTL models operate with expected demand (perhaps called estimated demand), or something slightly larger to account for normal variations, and at the planning level assume that all demand must be met.

Or to put it differently, these demands are what we set up the service network for. The modelers realize, of course, that in real time, orders will be different, and rejections might happen.

Hence, the modeling framework is that of planning to deliver the expected demand (or something slightly larger) but realizing that, while running the system, you might in fact reject some orders. When formulating a service network design model with random demand, you might have to rethink this strategy. Is it really correct to assume that you can carry it all? If you set up a two-stage stochastic program, with demand being stochastic and where the first stage is to set schedules (and determine the number of trucks) and the second stage is to send the packages, should you then require that demand be met with your trucks in all scenarios?

Let us see what that would imply. If the demand you use in your model is based on history, for example, by simply using observed demand as scenarios (see Sect. 4.1.2 for a discussion), it should be obvious (right?) that the most extreme (high demand) scenarios will drive the solution. You might have to hire a full truck to transport one package in one single scenario (and that is all you use the truck for throughout a week, and only if that scenario occurs). That is clearly unreasonable in most interpretations of the model.

If you use all potential demand to estimate scenarios, an approach of being able to send anything is obviously unreasonable, so we will not discuss that any further.

So it seems the only reasonable model is to allow rejections in the model itself, that is, to plan to reject certain demands. Many will react negatively to explicitly planning to reject orders—because it might, for example, create bad publicity. Even so, the fact is that, also in the deterministic modeling framework, we planned to reject in real time. It is just that it did not show in the tactical model.

But what stochastic demand should we use? There is no right answer here. A right answer only exists if you *know* the future, stochastically speaking, and you do not! That is a strong statement, so reflect on it for a while. Except for such problems as playing at a casino (where the rules are known, and we hope the deck of cards is proper) can we know the future in terms of distributions. In all other cases, we simply hope that we know the future. And we cannot test if our view on the future is good or bad! We can test if the past was a good description of the future in the past. But that is as far as we can get.

> We hardly ever actually know the relevant distributions, as our interest is the future, and we do not know (and cannot test) if the past describes the future.

So far we have discussed two possible distributional assumptions: that the future is described by our own past and that we use all potentially available demand to estimate our scenarios. We hope you see fairly quickly that the latter is not a good idea! It would overestimate the demands available to us.

If this were a real problem for you to solve, finding the correct distributional assumptions would be hard. In fact, it might turn out to be the hardest part of the modeling exercise. We will now leave that behind and simply assume that we have a distribution describing our best estimate of future (random) demand, and we will assume that packages can be rejected. But we hope this discussion has shown that the issues involved are complicated, with respect to both distributions and whether or not a constraint should be hard or soft (i.e., should be enforced or deviations penalized).

5.4 How We Started Out

The preceding discussion was far from obvious to us when we started out. First, we started out with demand scenarios that represented what we *had* to transport (so no rejection—in line with the deterministic models) and the scenario-generation method of Sect. 4.4.2. We were unable to obtain in-sample stability as defined in Sect. 4.2. So we asked why. The reason was, in our view, an interesting effect of two phenomena.

1. The scenario-generation method of Sect. 4.4.2 makes trees that match four marginal moments and correlations. It also keeps the approximate shape of the initial marginal distributions put into the iterative procedure of the method. But—and this turned out to be important—the extreme values were not exactly the same in all scenario trees.
2. In an integer program of this type, an arbitrarily small increase in the maximal possible demand can result in the need for an extra truck and, hence, a jump in costs. Alternatively, it can result in substantially different schedules (but the same number of trucks), also resulting in jumps in costs.

So we were concerned about instability. This concern is not only about a lack of numerical stability (and hence ability to solve) but is also a modeling concern. What is it with our model that makes it jump around like this? Remember what we pointed out previously, that scenario-tree generation is part of our modeling, as is the algebraic description of the model. A good model, in this extended sense, would be stable.

So our first step was to extend the method to handle bounded support, i.e., guarantee that no outcomes were generated outside the support. That turned out to be feasible, and we observed in-sample stability. But we were still concerned. A close look at the solution showed that some of the trucks were hardly used at all; they were only needed to transport packages with very small probabilities of showing up. And we had to ask: Is that how such companies are run? The answer was obviously no. That made us switch to a model with rejections. Which rejection cost to choose is of course a question in its own right. If rejection means sending packages with a competitor, then the rejection cost could simply be the extra cost of using the competitor. If, in addition, there is concern about reputation, it may be set a bit higher. If rejection is literally a rejection, lost customers and reputation must both be included. Rejection may also refer to your sending the packages yourself but by a mode not covered by the model.

Out of these stability tests came a revision of both the algebraic model and the modeling of scenarios. It is worth repeating that in-sample stability is not only a numerical issue; it is also a modeling issue. A model that is not in-sample stable is not good.

5.5 The Stage Structure

As it stands, this problem is inherently two-stage, with inherent Stage 1 being the setup of the network and inherent Stage 2 the use of the network. The number of actual stages in the problem is large (possibly infinitely large), but we do not want to model that. In fact, having infinitely many stages in a transient model will not do. On the other hand, our interest is the design, not the dispatch. We will use that to arrive at a two-stage model. It is worth noting what is going on here. The way we will model commodity flow will represent the flow well enough to obtain a good design: the model will "understand" that the design must function under many different demands. But the commodity flows will *not* represent possible dispatches in reality, as they do not capture the dynamics of demand realizations.

> In inherently two-stage problems we are not really interested in the variables belonging to the inherent second stage. Hence, we can simplify as long as the effects on the first-stage variables are kept.

However, we cannot remove the variables, as that would remove the signal for how to obtain good robust schedules. Many inherently two-stage problems can be modeled this way—by realizing that the variables of the inherent second stage are not really needed for implementation.

5.6 A Simple Service Network Design Case

The proposed model starts from a deterministic fixed-cost, capacitated, multicommodity network design model. Integer-valued decision variables are used to represent service selection decisions, whereas product-specific continuous variables capture the commodity flows. For instances, where the service offered is not able to meet the demands, we introduce a penalty cost (or an outsourcing cost). The penalty can refer to plain rejection, the use of a competitor, or possibly sending the commodity with another service offered by the company itself, be that a more expensive or slower service. The goal is to minimize the total system cost under constraints enforcing demand, service, and operation rules and goals.

Several simplifying assumptions are made:

- We consider a homogeneous fleet of capacitated vehicles with no restrictions on how many vehicles are used.
- The transport movements require one period, whereas terminal operations are instantaneous (within the period).
- Demand cannot be delivered later than the due date but may arrive earlier.
- There is a (fixed) cost associated with operating a vehicle (service), but no cost is associated with moving freight (except when outsourcing is used), that is, truck movements cost the same whether the trucks move loaded or empty.
- There are no costs associated with time delays or terminal operations.
- The plan is repeated periodically.

The space-time network is built by repeating the set of nodes (terminals) $\mathcal{N}$ in each of the periods $t = 1, \ldots, T$. Each arc (i, j) represents either a service, if $i \neq j$, or a holding activity, if $i = j$. A cost c_{ij} is associated to each arc (i, j), equal to the cost of driving a truck from terminal i to j if $i \neq j$, or to the cost of holding a truck at terminal i if $i = j$. The cost of outsourcing one unit or a failed delivery of a unit is represented by b. A period-by-period complete network is assumed. For each commodity $k \in \mathcal{K}$, we define its random demand $\delta(k)$, origin $o(k)$, destination $d(k)$, and the periods $s(k)$ and $\sigma(k)$ when it becomes available at its origin and must be delivered (at the latest) at its destination, respectively. The truck capacity is denoted M.

Stochastics are described in terms of scenarios $s \in \mathcal{S}$. To each scenario is attached a probability $p^s \geq 0$, with $\sum_{s \in \mathcal{S}} p^s = 1$. A scenario is $\mathcal{K}$-dimensional, as it contains one demand for each commodity. To indicate the impact of demand variation, the Y and Z flow variables are now indexed by s. The demand for commodity k in scenario s is given by $\delta(k, s)$.

In any multiperiod formulation, one must address end-of-horizon effects. We can mitigate this problem by casting the model in a circular fashion. Assuming a T-period planning horizon, a circular notation means that the period preceding period t is given by

$$t \ominus 1 = \begin{cases} t - 1 & \text{if } t > 1, \\ T & \text{if } t = 1. \end{cases} \tag{5.1}$$

This approach is allowed since this is an inherently two-stage problem where our real interest is the design variables, not the flow variables. So we capture the effects of uncertainty, but the resulting dispatch is useless.

The decision variables and model are as follows:

$Y_{ij}^{t \ominus 1}(k, s)$: Amount of commodity k going from terminal i in period $t \ominus 1$ to terminal j in period t in scenario s.

$X_{ij}^{t \ominus 1}$: Number of trucks from terminal i in period $t \ominus 1$ to terminal j in period t.

$Z(k, s)$: Amount of commodity k sent to its destination using outsourcing or simply not delivered at all in scenario s.

Note that the circularity of the network means that we can have $\sigma(k) < s(k)$; if we have, for example, a commodity in a one-week network ($T = 7$) with $s(k) = 5$ and $\sigma(k) = 1$, that would mean that it becomes available on Friday (each Friday) and must be delivered by the following Monday. To facilitate this fact in mathematical formulations, we define a set of periods that a commodity can be shipped from:

$$\mathcal{F}_c = \{s(k), \dots, \sigma(k) - 1\} \text{ if } \sigma(k) > s(k) \text{ and } T \setminus \{\sigma(k), \dots s(k) - 1\} \text{ otherwise.} \tag{5.2}$$

The stochastic problem that we solve is

$$\min \sum_{ij \in \mathcal{N}} \sum_{t=1}^{T} c_{ij} X_{ij}^{t} + b \sum_{s} p^{s} \sum_{k \in K} Z(k, s); \tag{5.3a}$$

$$\text{s.t.} \sum_{i \in \mathcal{N}} X_{ij}^{t \ominus 1} = \sum_{i \in \mathcal{N}} X_{ji}^{t} \qquad j \in \mathcal{N}, \ t \in T; \tag{5.3b}$$

$$\sum_{i \in \mathcal{N}} Y_{ij}^{t \ominus 1}(k, s) - \sum_{i \in \mathcal{N}} Y_{ji}^{t}(k, s) = \begin{cases} \delta(k, s) - Z(k, s) & \text{if } j = d(k) \text{ and } t = \sigma(k), \\ -\delta(k, s) + Z(k, s) & \text{if } j = o(k) \text{ and } t = s(k), \\ 0 & \text{otherwise;} \end{cases}$$

$$j \in \mathcal{N}, \ t \in T, \ k \in K, \ s \in \mathcal{S}; \tag{5.3c}$$

$$\sum_{k \in K} Y_{ij}^{t}(k, s) \leq M X_{ij}^{t}, \qquad ij \in \mathcal{N}, \ i \neq j, \ t \in T, \ s \in \mathcal{S}; \tag{5.3d}$$

$$0 \leq Z(k, s) \leq \delta(k, s), \qquad k \in K, \ s \in \mathcal{S}; \tag{5.3e}$$

$$X_{ij}^{t} \geq 0 \text{ and integer}, \qquad ij \in \mathcal{N}, \ t \in T; \tag{5.3f}$$

$$0 \leq Y_{ij}^{t}(k, s) \leq \delta(k, s), \qquad k \in K, \ ij \in \mathcal{N}, \ t \in T, \ s \in \mathcal{S}; \tag{5.3g}$$

$$Y_{ij}^{t}(k, s) = 0 \text{ for } t \notin \mathcal{F}_k, \qquad k \in K, \ ij \in \mathcal{N}, \ s \in \mathcal{S}. \tag{5.3h}$$

Objective function (5.3a) minimizes the sum of the first-stage costs for opening the arcs plus the expected rejection costs. Constraints (5.3b) enforce the conservation of flow for trucks; the "$t \ominus 1$" expression in the subscript ensures that the trucks form circular routes. Constraints (5.3c) play a similar role for the commodity flow, except that the flow has an origin and a destination instead of being circular. The volume of the flow is equal to the satisfied demand, i.e., the difference between the scenario demand $\delta(k, s)$ and the rejected demand $Z(k, s)$. Constraints (5.3d) limit the total flow on each arc to the booked volume on the trucks, and constraints (5.3e) limit the rejected demand to the actual scenario demand. Note that (5.3d) do not apply on holding arcs with $i = j$ since we assume that commodities can be held at the nodes without a truck being present and with no capacity restriction.

Finally, constraints (5.3h) ensure that the path for commodity c does not loop around the network and thus prevent deliveries from taking more than T time periods. We could also set to zero all the Y variables in the origin period $o(k)$ that do not start from the origin node $d(k)$ and do the same for the destination, but that would make the formulation more difficult to read; in addition, this can be expected to be taken care of by an LP presolver.

When this model is solved, we obtain the number of trucks to use, their fixed routes, the flow of commodities (one set of flows per scenario), the amounts outsourced (also by scenario), and, finally, the total expected cost. Figure 5.1 shows a typical example of truck routes from the model.

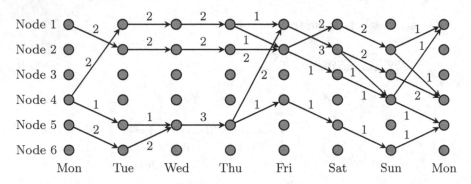

Fig. 5.1: Typical first-stage output from model. The first and last columns represent the *same* day. Numbers on arcs refer to the number of trucks running in parallel on the arcs

It is worth noting the interpretation of the flows of commodities. We obtain one set of flows per scenario. Hence, when flow demands actually arrive in sequence (and not scenario by scenario), we cannot use the flows directly for routing goods. However, we chose this approach as it helps us achieve good robust schedules, avoids end-of-horizon effects, and at least provides guidance for flow.

In an inherently two-stage model we are not really interested in decisions in later stages, but we need to send good signals to the first stage. This allows up for useful approximations.

5.7 Correlations: Do They Matter?

As an example of how ideas from this book can be used to study a new problem, let us ask ourselves the following question: How important are correlations? Does it matter if we get them right or not? We must remember that deterministic models have nothing that corresponds to correlations. So if they are important, we need a tool that can relate to them. We outlined this in Sect. 1.6.2.

To study correlations, a scenario-generation procedure allowing us to actually control correlations is needed. We used a slightly updated version of the method in [28]. That method allows us to control the first four moments of the marginal distributions, plus the correlations. The extension ensures that we stay within the support of the random variables, as we have already explained. It should be clear that, had we used sampling to obtain scenarios, we could not have controlled the correlations (or the marginal distributions for that matter) unless we had created very large trees. But with large trees, the model would not be numerically solvable.

With this approach we obtained in- and out-of-sample stability as outlined in Sect. 4.2. Let us see where this takes us. With an in-sample stable scenario tree and an optimization problem of limited size, we can investigate our main concern: the importance of correlations for the optimal solution.

5.7.1 Analyzing the Results

In this example, demand is described by triangular distributions, each defined by its min, mode, and max. In addition, we have a correlation matrix. Due to the complexity of the problem, and for a better overview of the results, we limit ourselves to using no more than 16 random variables and 252 integer variables.

5.7.1.1 Flexible or Robust?

Let us first reflect on flexibility and robustness. What do we actually want to achieve in this case? The schedules, as seen by the customers, should be robust, that is, it should not be necessary to change them in the light of shocks. This aspect is not really part of our model as we, by defining the commodities, have required this robustness to be present. In our model, there is no possibility of dropping a service. It is feasible to outsource a full service, but from a customer's perspective, the service is still there. The services, as seen by customers, refer to what services are offered, not how the goods are transported. So this aspect is part of the input to the model and is taken care of that way.

Second, by *requiring* the model to have schedules for the trucks, we also immediately achieve robustness for drivers and trucks. That is, the varying demand does not result in repeatedly new plans for the trucks and drivers. We might have defined models where trucks could be rerouted, but we did not. So the resulting schedules are robust, as required, from the drivers' perspective.

5.7.1.2 Flexibility in Routing

The correlation matrices used in our experiments were created to represent three structurally different situations (Fig. 5.2). The actual matrices will obviously be a bit larger than what is shown here. The three cases are those of uncorrelated random variables, strongly positively correlated variables, and a case where strong positive and negative correlations are mixed. Uncorrelated demand implies that there are no systematic relationships among the demands from our customers. Strongly positive correlations imply that there are some underlying phenomena that drive the demands of our customers, mostly making them all large or all small (although there is a small probability that one is small while another is large). The mixed case represents a situation where we have several sets of customers. Within each set, correlations are strongly positive, while between sets, the correlations are negative. A simple example could be that we have a set of customers experiencing large demands when the weather is good and another experiencing large demands when the weather is bad. But on top of these trends, there are individual variations, hence the correlations are not ±1. Negative correlations also occur if a given customer has two alternative production sites (origins), and we know for sure that he will send goods in a given week, but we know neither from which origin nor how much. Also, if a customer is known to send goods in a given week, but the day is uncertain, these two origin–destination pairs will have negatively correlated demand (as they will be seen as different commodities in the model).

Note that the case of all correlations being negative does not exist in any interesting way. Three demands cannot all be pairwise negatively correlated unless the correlations are rather small.

0 Correlation	Mixed Case	0.8 Correlation
1 0 0 0	1 0.8 -0.8 -0.8	1 0.8 0.8 0.8
0 1 0 0	0.8 1 -0.8 -0.8	0.8 1 0.8 0.8
0 0 1 0	-0.8 -0.8 1 0.8	0.8 0.8 1 0.8
0 0 0 1	-0.8 -0.8 0.8 1	0.8 0.8 0.8 1

Fig. 5.2: Structure of the three correlation matrices used in the tests

In the deterministic case below, we simply use expected demand. The major results are summed up in Table 5.1.

As we can see from Table 5.1, four trucks are needed in all cases. In two cases, even the operating costs are the same. However, this does not imply that the trucks follow the same routes. In fact, the truck routes, as well as the flows for the commodities, are rather different, and that is why the solutions behave very differently, as we will soon see.

Table 5.1: Results of four cases solved

Case	No. of trucks used	Cost of operating the schedule	Expected cost of outsourcing	Total cost
0 Correlation	4	4,850	396	5,246
Mixed case	4	4,850	285	5,135
0.8 Correlation	4	4,600	565	5,165
Deterministic	4	4,700	0	4,700

The total costs are given in the rightmost column. All the stochastic cases are rather similar. The deterministic cost is quite a bit lower than the three stochastic cases. However, as this is the optimal objective function value within the optimization problem and not the true expected cost of using the solution from the deterministic model, this low cost is misleading (as it normally is in deterministic models). We will return to the true expected cost in a short while. Note that outsourcing was allowed also in the deterministic case.

5.7.1.3 Use of Outsourcing

Although the total expected costs are similar in the three stochastic cases, the use of outsourcing is somewhat different. With strongly positive correlations we are a bit modest while setting up the network and rather choose to plan for a somewhat large amount of outsourcing.

5.7.1.4 Strong Positive Correlations

There are clear differences in the needs in the three stochastic cases. For strongly positive correlations, the probability of all (most) demands being large at the same time is rather high. There will be a few scenarios that really drive the solution. The focus in this case is on facilitating these cases of high demand, and the result is a plan where parts of *really* high demand scenarios are foreseen as being outsourced, but where most other scenarios are easily accommodated within the given solution. Low costs here come from the fact that the schedule is set up to accommodate the most probable scenarios, and these are scenarios where most demands are high at the same time. Money is not spent creating alternative connections in the network. Rather, one sets up a network that is well suited for the most likely high-demand scenarios, and money is otherwise spent on carefully planned outsourcing.

The need to be flexible in the routing of goods in the network is low, as we are quite close to a worst-case (stochastic) situation, given by all correlations' being equal to one. Hence, flexibility has taken a back seat relative to the need to accommodate the high-probability, high-flow scenarios.

5.7.1.5 Mixed and Uncorrelated Cases

In the other two cases, we are much farther away from the worst case, and we need a network that can facilitate a wide variety of rather different flows. When scenarios with some demands high and some low have high probabilities, we need flexibility in the possible ways of routing flow. The reason is that there are many different ways of combining high and low demands, none of them very likely, but the total probability of such combinations is very high. Hence, we invest in flexibility in routing. One way of achieving this flexibility is to use consolidation. When two negatively correlated demands can share a path, we easily achieve a very effective use of capacities if these two flows are consolidated. Hence, we here invest more in routing flexibility and obtain much less outsourcing. Zero correlations also allows for a certain amount of capacity sharing, although less than when we have negative correlations.

5.7.1.6 What If We Are Wrong About the Correlations?

As important as looking at how costs are distributed is looking at how the solutions perform under conditions different from those for which they were originally intended. We take the schedule obtained from solving our problem using a scenario tree based on one correlation matrix and test it using another scenario tree based on the same marginal distributions, but now with a different correlation matrix. To do this test, we simply fix the network design variables in our general model and solve only for the second stage. The results are found in Table 5.2.

Looking at the columns in Table 5.2 we note something interesting. The performance of the various solutions are very different when they are tested under conditions they were not intended for. Why is that? We mentioned earlier that for the case of strong positive correlations much of the focus of the design is on facilitating the really difficult (high-demand) cases (as they have high probabilities). A side effect of this is that the schedule is not very flexible with respect to the routing of goods. Hence, when scenarios it was thought had very low probabilities (and which were planned to have rather much outsourcing) turn out to be rather probable, the expected outsourcing costs become quite high. The mixed case knows that many peculiar scenarios can occur. As a result, the design allows for more flexible in routing, and hence it handles the other cases much better. The zero-correlation case has problems with strongly positively correlated demands as it does not handle the high-demand cases well (and they now have much higher probabilities than planned for).

Table 5.2: Expected costs resulting from using four different solutions in three different stochastic environments

Case	Expected cost of outsourcing		
	0 Correlation	Mixed case	0.8 Correlation
0 Correlation	396 (396)	290 (290)	2,141 (2,141)
Mixed case	432 (432)	285 (285)	842 (842)
0.8 Correlation	1,802 (1,552)	1,776 (1,526)	565 (315)
Deterministic case	973 (823)	681 (531)	1,402 (1,252)

The numbers in parenthesis are adjusted for the differences in design costs for the networks

5.7.1.7 The Most Flexible Setup

Hence, we see that getting the correlations wrong when planning can lead to serious problems later on. We also see that the deterministic case has its difficulties, and in all cases it is the second worst. For this example we see that the mixed case behaves best on average and has the best worst- case behavior. This is probably caused by its producing a network for handling all types of structures, resulting in a desirable flexibility in routing.

5.7.2 Relation to Options Theory

We may choose to see these results in light of options and option values. As can be seen from Table 5.1, the first two schedules cost the same. Of course, when their assumptions of correlation structures are correct, they are optimal. But consider how they function for the case of strongly positive correlations in Table 5.2. One has a much better expected performance than the other. Hence, it seems that the mixed case has produced options with much higher values than those coming from the zero-correlation case. The actual value of these options will, of course, depend on what is the correct situation. If 0.8 correlations is the correct description, then the options from the mixed case are worth $2,141 - 842 = 1,299$ more than those of the zero-correlation case. It is worth noting that since we do not know what the options are, option theory—as it is normally defined—cannot help us find these values. Option theory can only value options that are predefined. Stochastic programming gives us the values directly and, by providing us with optimal decisions, at least can help us try to understand what they are. We will return to that issue in a few moments.

5.7.3 Bidding for a Job

A major observation from our case is that there certainly exist situations where it is crucial to get the correlation structure right. We also see that if we get it wrong, some settings are still better than others. Flexibility and robustness are important issues here. Some choices of correlations lead to better solutions with respect to the flexibility in the routing of goods. Understanding this process is crucial for practical use. Overlooking correlations totally by resorting to deterministic models is equally dangerous. Based on our findings, it is crucial for an LTL carrier or an airline to correctly estimate covariation, and it is certainly important to take stochastics into account. Otherwise, a chosen schedule may turn out to be very fragile.

Consider some carriers, independent of mode, who participate in a competitive tender. Assume we have rational participants knowing that uncertainty exists (and who therefore use stochastic models) and that the participants do not face any form of economy of scale/scope. If so, the players would try to win the tender by obtaining the largest possible profit margin. Assume further that these players use the same planning tools and have the same objective (profit) function, so that the only difference between them would be how they perceive the "world." Narrowing it further down so that the only difference between the participants is how they think the demands are correlated, we could still end up with substantially different bids. Using the results from this section we can clearly observe that a firm believing that the correlations were like the mixed case would win the bid (assuming identical planned profit margins). The only problem is that if the demands turned out to be completely uncorrelated or strongly positively correlated, winners curse would be likely to occur. The winner would lose money. A carrier aware of this problem could of course make a bid based on a worst-case view on the correlation matrix. That would result in a bid not suffering from winner's curse, but also a losing bid.

5.8 The Implicit Options

So what is going on when some schedules provide more flexibility in routing than others? What are the options inherit in the solutions? What does a good robust schedule look like? To find that, we must study the solutions we found. Even though the solutions obviously contain different options, they are hard to find. This is important to realize. Option theory focuses on well-defined options, like the option to wait or the right to do something specific in the future. Here it is more subtle. You cannot subtract a nonrobust solution from a robust one and interpret the difference as an option. The solutions are simply different, and one is better than the other. So, in a sense, it is easy to find option values (and costs) but not to actually find the options. This marks the difference between stochastic programming and option theory. Option theory can only value structures already defined; it cannot find them. Stochastic programming can value total solutions, obviously containing options, but finding what the options are must be done manually. Even so, this puts us in a much better situation than if we had just option theory to work with. We will now see that for the service network design problem, we have indeed found out (at least to some extent) what characterizes a good robust schedule with flexible routing of goods.

> Options are implicit in the solutions, not explicit on top of another solution.

We will see that, on the one hand, the structures are somewhat simple, whereas on the other hand, deterministic models would not produce these structures. We will also observe that the well-known hub-and-spoke structure, in particular the idea of consolidation, will be part of the solution structures without being enforced. It is simply optimal, in light of randomness in demand, to use consolidation. Under certain reasonable conditions, consolidation takes place in a hub-and-spoke environment.

5.8.1 Reducing Risk Using Consolidation

Consolidation in LTL trucking is normally thought of as a way to accommodate the fact that most loads are less than one truckload. In deterministic models, consolidation helps keep truck utilization up and also facilitates the use of reasonably large trucks over longer distances. However, there are also risk-related reasons for using consolidation. By risk in this case we simply mean the risk of having to reject or outsource demand as that (presumably) is bad for business.

Let us look at a minor example where each load has an expected size equal to the truck capacity. In such a setting, consolidation will not be part of a deterministic solution. Consider the example with two commodities in Figs. 5.3 and 5.4. One commodity becomes available at node 1 at time 0 and has to be delivered to node 2 within three time periods. The other commodity becomes available at node 3 at time 1 and has to be delivered to node 2 within one time period. The thinner arrows in the two figures show the optimal schedule for the two trucks that will be used to transport the commodities from their origins to their destinations. We note that there will not be any consolidation in the deterministic solution (assuming just these two commodities), so that each commodity will use its own truck. Note that the solutions are not unique (but all solutions have the same structure).

In the stochastic solution, we see both trucks following the same path. This is a different and, in the example, slightly more expensive schedule, compared to the deterministic one, as the trucks move more. However, it allows the two commodities to share the joint capacity of the two trucks from node 3 at time 1 to node 2 one period later. This is an example of where commodity 1 is sent trough an intermediary breakbulk (node 3) before being consolidated with commodity 2 using the two trucks servicing the link from node 3 at time 1 to node 2 one period later. Having this kind of solution, where two commodities share the common capacities of two trucks, makes expensive outsourcing less likely compared to if one truck had been used by each company as in the deterministic case.

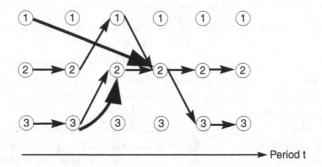

Fig. 5.3: Deterministic problem with no consolidation, using two trucks

Let us illustrate using some numbers. Assume trucks have capacity 2, and let the demand for each of the commodities be 0, 2, or 4, each with a probability of one-third. Hence, in total, there are nine possible scenarios, and if the demands are uncorrelated, each scenario has a probability of one-ninth.

5.8.1.1 Hedging Using Consolidation

Let us see what happens if we use the deterministic solution of Fig. 5.3 but in fact face this uncertainty. Whatever demand cannot be met will be outsourced. Of the nine scenarios, five will result in outsourcing, and the expected amount outsourced is four-thirds. If we test the schedule in Fig. 5.4, we note that the

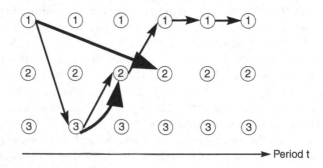

Fig. 5.4: Stochastic problem using consolidation between nodes 3 and 2 (two trucks used)

expected amount of outsourcing will be eight-ninths, or four-ninths lower than for the deterministic schedule. What has happened is that the two scenarios $(4, 0)$ and $(0, 4)$ can now be handled without outsourcing due to consolidation, that is, due to the ability to share transportation capacity.

What we observe here is closely related to what has been observed in other parts of the operations research literature. For example, in inventory theory, when the number of warehouses/inventories drops, safety stocks can be reduced without reducing the service levels. In finance, the risk in a portfolio of various financial instruments can be reduced by diversification, keeping the expected return the same.

Such use of consolidation as a means to hedge against uncertainty is a feature that has not been described in the service network design literature. There can be several reasons for this, but since the literature almost exclusively refers to or uses deterministic models, hedging against uncertainty becomes irrelevant. This is so despite the fact that most researchers in the field are well aware that uncertainty plays a major role in problems. The models used, however, are deterministic. Notice again that consolidation is not a property we enforce on the solution but a structure that emerges because it is good for the overall behavior of the schedules.

5.8.2 Obtaining Flexibility by Sharing Paths

In the foregoing example there are only two commodities. Let us now pass to an example with four commodities. We will see that it is good not only to consolidate but to have many paths available for each commodity as well. In addition, all observed aspects of consolidation here will come from stochastics and not from standard volume-related arguments.

Figure 5.5 illustrates an example. We have four commodities that become available at their respective origins in the first time period and have to be transported to their destinations within the time indicated by the arrows. This transport can take place either by using the company's own trucks or by outsourcing. The schedule in Fig. 5.6 is based on stochastic demand, while for the schedule of Fig. 5.7, demand equals the expected demand from the stochastic case. Otherwise, all parameters are the same.

The deterministic and stochastic schedules require the same number of trucks. In the deterministic case, no outsourcing is used (when the demand is

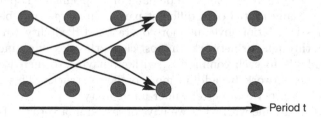

Period t

Fig. 5.5: Origin and destination of commodities

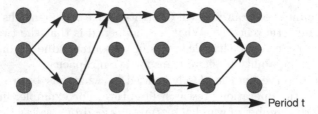

Fig. 5.6: Schedule of trucks based on stochastic demand

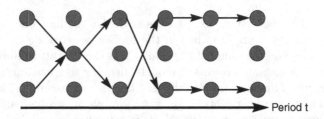

Fig. 5.7: Schedule of trucks based on deterministic demand

deterministic), whereas the stochastic case uses some outsourcing (less than 0.1% of the total flow in all the scenarios) to be able to transport all the commodities to their destinations. (This does not imply, however, that the deterministic solution is better with respect to outsourcing. When the deterministic solution is subjected to random demand, a total of 1.17% of the total flow is outsourced.)

5.8.2.1 More Flexibility in Routing for Stochastic Case

The most substantial difference between the two solutions is the number of paths connecting the various O–D pairs. In the deterministic case, there are two O–D pairs that are connected with one path, while the two others are connected by two paths. In the stochastic case, each O–D pair is connected by at least two paths. The higher number of paths in the stochastic solution makes it easier to switch the flow of commodities from one path to another if required because the first path is taken by another O–D pair. Such situations might occur when there is a surge in demand on a specific O–D pair that can be fully or partially routed on a different path. This larger number of paths in the stochastic solution gives us more operational flexibility when routing commodities through our network. In most cases, this flexibility increases the capacity available to each commodity, without having to increase the total capacity in the network by adding more or larger trucks. This makes our stochastic solution perform better under uncertainty.

This result is in line with what we have observed in a variety of test cases. When uncertainty becomes important, our solutions habitually move away from "direct connections" between origins and destinations in deterministic cases to more hub-and-spoke-looking networks where the freight is being shipped trough intermediary terminals. This is not because it was "decided" a priori that freight between some O–D pairs should be handled this way, but because it turned out to be the best solution to deal with the uncertainty. Using our model has shown us that consolidation in hub-and-spoke networks takes place not necessarily due to economies of scale or other similar volume-related reasons but as a result of the need to hedge against uncertainty.

So the conclusion so far is that a good solution will try to provide many paths for each O–D pair such that capacity is shared with other (different) O–D pairs on each of these paths. This provides optimal operational flexibility. It implies that as soon as some demand is low, others can immediately utilize the freed capacity if need be. A deterministic model would never produce such structures.

5.8.3 How Correlations Can Affect Schedules

Look back at Figs. 5.3 and 5.4. When we calculated the value of using the stochastic rather than deterministic model, we assumed that the demands were independent. Assume instead that the two demands are perfectly negatively correlated. In that case, the sum of the two demands will always be 4, and the expected outsourcing will be zero. But if we use the solution from the deterministic model, expected outsourcing will remain at four-thirds. What we observe is not surprising but important. Negative correlations imply a chance to achieve hedging, but only if the two negatively correlated flows can be set up so as to share capacity. And the more negatively correlated they are, the more important the issue.

Always look for negative correlations.

On the other hand, in the same example, if the two demands are perfectly positively correlated, the expected outsourcing will be the same for the deterministic and stochastic solutions since, in fact, consolidation will never take place.

More generally, as long as random variables are not perfectly positively correlated, there is something to be gained from flexible routing. The more we move toward perfectly negative correlations, the higher is the potential for hedging. As flexibility normally comes at a cost, there is a trade-off between the cost of achieving operational flexibility and the expected gain from the investment. Of course, when there are many random variables, the relationships are more complex. But even so, we can conclude as follows: operational flexibility is achieved by having many paths available for each O–D pair such that each of these paths is shared by other O–D pairs. The more negative the correlations are, the more there potentially is to be gained from well-structured schedules.

5.9 Conclusion

The purpose of this chapter has been to illustrate how the ideas in this book can be used to analyze a new problem. The setting here is that of a graduate student studying the effects of uncertainty on a discrete optimization problem. We trust other students can do the same with other problems. Note the need to use proper scenarios from Chap. 4 to make sure that you analyze what you think you are analyzing.

Chapter 6

A Multidimensional Newsboy Problem with Substitution: With Hajnalka Vaagen

I suffer from a severe fashion disorder.
– *Cosmo Fishhawk*

Fashion is a form of ugliness so intolerable that we have to alter it every six months.
– *Oscar Wilde*

In this chapter, you will encounter a problem that is inherently two-stage. The first inherent stage is to decide on production levels for a number of substitutable products with correlated demands. This is followed by a second stage where demand is met, partly by giving customers what they want, partly by giving them acceptable substitutes. The chapter is based on [53].

You will see that the model, even if used in a full-scale industrial setting, is small, with negligible CPU time. However, this does not mean that the problem is straightforward. The modeling itself is original, and, maybe most importantly, the distributions are complicated. So to solve this problem satisfactorily, we will need to call upon results from Sect. 4.4.2. Not all stochastic programs are hard because they are large. Complexity may also come from modeling as such and from complicated distributions. We will see how simplifying assumptions on the distributions will cause severe losses of profit.

> Stochastic programming offers the ability to handle complicated distributions. Difficulties do not always stem from size.

6.1 The Newsboy Problem

Let us first review the classical newsboy problem. It is an interesting stochastic optimization problem in its own right. Imagine a boy selling newspapers at

A.J. King and S.W. Wallace, *Modeling with Stochastic Programming*,
Springer Series in ORFE, DOI 10.1007/978-0-387-87817-1_6,
© Springer Science+Business Media New York 2012

a street corner in a big city. Each morning he goes over to the newspaper headquarters to buy a number of newspapers at a unit cost of c. He then goes to his street corner. For each paper he sells he gets v, while each paper not sold gives him a salvage value (maybe recycling) of g. He has been selling for a long time, so he has a good feeling for the demand, and it can be described by a scenario set d^s for $s \in S$. So in this problem, he can order newspapers only one time. That is a crucial aspect of the newsboy problem.

> The newsboy problem is about perishable products where all production must take place before demand becomes known.

Let

x be the number of newspapers bought,
y^s be the number of newspapers sold in scenario s,
w^s be the number of newspapers not sold in scenario s,
c be the purchasing price of a newspaper,
v be the selling price, and
g be the salvage value for an unsold newspaper.

The problem is then

$$\max \quad \sum_{s \in S} p^s(vy^s + gw^s) - cx$$

$$\text{such that} \quad \begin{cases} y^s \leq d^s, & \forall s \in S, \\ w^s = x - y^s, & \forall s \in S, \\ x, y^s, w^s, \geq 0 & \forall s \in S. \end{cases} \tag{6.1}$$

Note that $y^s \leq x$ is implied by the other constraints.

This problem has an analytical solution, but we will not worry about that now. It can also be solved numerically, of course, and the extension presented in this chapter can only be solved numerically.

The problem we are facing in this chapter extends the classical newsboy model in two directions. First, we have several *items*, as we will call them, not just one. These will all represent the same basic need for the customer. Think of things such as high-tech mountaineering jackets for men (in different colors and designs). These are typically fashionable in only one season, and the production lead times are so large that all production decisions must be made before demand is realized. The salvage value corresponds to selling on sale or throwing away. Since these items satisfy the same basic need, the demand for them is strongly dependent. We will return to how to describe the demand distributions.

Note that this is a newsboylike problem by *construction*. The products are made at specialized facilities in China and sent to the main markets by ship. Had the manufacturer decided to produce closer to the markets or sent the products by plane, we would not have faced a newsboylike planning problem. But that is not done as it would be too expensive.

The other extension in this model is that we will allow product substitution. That is, if you come in wanting a jacket of a certain color and design but can only get the right color but in another design, there is a certain chance that you will accept a substitute. And the pattern of substitution is as complex as the demand distributions, but we will return to that in a while as well.

The actual setting of the model is not that of a retailer and a customer like you, but that of a manufacturer and its retailers. In this context, substitution will imply that the manufacturer does not deliver to a retailer what he wants but an acceptable substitute or nothing at all. So only a certain percentage of those who are denied their first choice will accept a given substitute. This is called manufacturer-directed substitution and is different from consumer-directed substitution, which is much more difficult to model.

6.2 Introduction to the Actual Problem

Capturing market trends and satisfying customer demand by supplying quality products in a very short time is the dominant challenge in modern manufacturing. Sophisticated information technology enables fast and accurate information flow. However, when information available at the time of planning is largely qualitative and based on trend estimates and lead times are pressed to a minimum (but still very large), it is crucial to know what kind of information to look for, how to interpret it, and how to apply it in portfolio and production planning. Obviously, mapping the true problem complexity and focusing on the quantitative aspects are crucial for creating good solutions. Further, worried over the variety explosion in contemporary markets and its negative impact on their own supply chain activities, many suppliers and retailers turn from offering greater variety to a more efficient assortment strategy.

In this chapter, we discuss the quantitative aspects of assortment planning in quick response supply chains; sports apparel is taken as a case of analysis. Despite the fact that assortment planning is a rather new field, the motivation to study the problem is high, and much published work can be found; see [39, 42] for extensive reviews. Most of this work focuses on analytical formulations of assortment optimization. Different heuristics and strong assumptions on the nature of demand patterns and interitem dependencies are applied to achieve solutions. Discussions on the numerical complexity, tractability, and applicability of these formulations to real-life problems, as well as empirical tests of the theoretical predictions, are rather vague. The potentially enormous academic contribution in adding rigor and science to retailers' developed practices is emphasized in [39], much as it has been done in areas like finance. We will see that a numerical stochastic program can bring results not available via analytical models because of such a program's ability to handle complicated distributions.

6.3 Model Formulation and Parameter Estimation

Based on the recognition that multidimensional newsboy models with complex dependencies are analytically difficult to deal with, we analyze whether a numerically based stochastic programming formulation provides more useful solutions. We avoid using heuristics, thereby allowing for complex distributions, correlations, and substitution patterns. It is obvious that we are dependent on data, data that cannot be obtained without internal understanding of the specific industry. However, we do not consider this to be a disadvantage but rather recognition and acceptance of reality.

We assume cost and selling prices are rather homogeneous within the group of substitution. The trend-driver attributes, on the other hand, are qualitative and heterogeneous. When demand becomes known, products are assigned to customers so as to maximize the manufacturer's profit. Optimization happens in one stage, with allocation between direct and substitution sales. The optimal solution implies, then, that the manufacturer controls these values; in other words, it tells the retailers how much of the first and lower preferences they may buy. For a real-life situation with characteristics similar to this problem, we refer to the fashion and sports apparel franchises.

6.3.1 Demand Distributions

When the major decisions have to be made, that is, which items to include in the portfolio, and how much to produce, only limited information is available about demand. The reason is that we are facing a product affected by fashion, and it is not known what will be "hot" in the coming season. However, much is known about the structure of demand. Some items are known to be in competition with each other. For example, it might be known that if navy is a popular color, then black will not be popular. Or it could be the other way around, with black popular and navy not. This will be expressed by negative correlations in demand. Positive correlations will occur for colors that are popular (or unpopular) together. In one season, all clear colors may be popular, while no earth colors are. So correlations can be estimated on the basis of such knowledge. Some items will be more stable—such as white—and not so much affected by fashion. The same type of arguments can be made about design.

But there is more to it than that. Assume we have three groups of items. The items in Group A are popular with a certain probability, while those in C are popular when those in A are not. We say that the world is in State 1 if those in A are popular, and in State 2 otherwise. We also have a set of items that have a stable demand. We call this Group B. Figure 6.1 shows typical joint distributions for pairs of these groups for a case where the probability of each state of the world is the same, and all scenarios shown have the same probability. In the first panel, State 1 is the cluster of relatively high demand from both group A and B in the upper right corner, and State 2 is the cluster in the lower right quadrant representing relatively high demand from group

B and low demand from group A. The center panel shows the crucial group A and group C tradeoff, with the State 1 cluster in the upper left consisting of relatively high group A demand and low group C demand, and the State 2 cluster in the lower right depicting the low demand from group A and high demand from group C. The third panel shows the State 1 cluster in the lower right, with low group C demand and high group B demand, and the State 2 cluster in the upper right with high demand from both group B and group C.

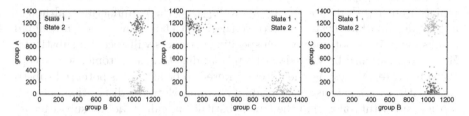

Fig. 6.1: Joint distributions for pairs of item groups as a function of the state of the world

Note how each marginal distribution is bimodal and how the bimodality is distributed over only two quadrants when we study two items at a time. This kind of distribution is impossible to handle in analytical models. But as we will see in this chapter, to get good results, we must take the bimodality into account. It is also worth noting that *given* the state of the world, the distributions are not particularly complicated and can, in fact, be described well with simple log-normal distributions. (We like to avoid normal distributions since they have positive probabilities for negative demand.) We use tools from Sect. 4.4.2 to create these discrete distributions.

> Numerical models can handle complex distributions, even multimodal distributions, using the tools of Sect. 4.4.2.

To define state-conditioned distributions, we use aggregated demand data across whatever will become popular/unpopular. Hence, instead of approximating the overall distribution, we generate scenarios for each state of the world independently using scenario-generation tools from Sect. 4.4.2. The overall distribution is then built by connecting all the scenarios by state probabilities (probabilities that sum up to one). Under limited information, the two states occur with equal probabilities. This way, the uncertainty around the individual items' popularity is captured.

6.3.2 Estimating Correlation and Substitution

Correlation and substitution measures express the dependencies among the individual items. Assortment planning in our quick-response settings (fashion and sports apparel) largely happens when the information is limited to qualitative knowledge of the trend drivers and aggregated estimates.

Here is how you may think: the existence of two states of the world for some items is described by strong negative correlations in demand for all pairs of competing items. Similarities on trend-driver attributes, like the existence of specific technical features across items, define positive correlations. We connect correlations and substitutions by a common information base and describe product substitution based on the grade of similarity with regard to the trend-driver attributes. We define the decision-independent *a priori substitutability* $\alpha_{ij} \in [0, 1]$, indicating the portion of customers willing to replace item j with item i. We are assuming that the manufacturer, although able to decide which customers receive their first choice and which are offered a substitute, does not know which specific customers will accept a substitute. Hence, α_{ij} can also be viewed as the probability that a customer will accept the substitute. We assume the n items offered to be all others potential substitutes with heterogeneous substitutability values. Note the distinction between a priori substitutability and the true substitution, called *factual* substitution; this latter is a decision-dependent outcome of an optimization process constrained by unsatisfied demand and availability.

For the connection between correlation and substitutability values consider the following example. The supplier is to make assortment decisions on two identical models in colors black and navy. Assume the information available for decision making to be as follows:

(a) Color is a strong trend driver, and only black or navy will become popular; this can be described by a strong negative correlation between their demands (for example, -0.5).
(b) Due to the similarity with regard to design, if one becomes popular and faces stock-out, it can partially be substituted by the available one (say, with a substitutability of 0.2).

Given the "competition" between the products, it is unrealistic to assume high substitutability. Customers looking for the popular black will likely visit other stores/suppliers to get their preferred choice.

6.4 Stochastic Programming Formulation

We model the following process. In a first step, most appropriately in the design phase, the manufacturer defines the similarity among the products with regard to the trend drivers, hence, the substitutability matrix. It is normal that $\alpha_{ij} \neq \alpha_{ji}$. While the safe color white may be an acceptable substitute for the more trendy pink, people wanting white are not very likely to accept pink as they want a subdued look.

Next, based on knowledge of demand, the manufacturer describes marginal distributions, conditioned on the state of the world, and correlations. Finally, the manufacturer assesses the probabilities of the state of the world. If it knows nothing, the probabilities are set at 50% each.

Given substitutability and demand distributions, the manufacturer decides the optimal assortment to offer: products to include in the portfolio and their inventory levels. Finally, when the actual retailer demand becomes known, the manufacturer assigns first and substitute preferences so as to maximize expected assortment profit, given the initial inventory levels and substitutability matrix. The outcome of this process is the factual substitution.

A simple two-stage stochastic program is formulated. The first stage consists of the production decisions and the second stage (after demand has been observed) optimally allocates direct and substitution sales.

Sets:

S—set of demand scenarios
I—set of items in the reference group portfolio

Variables:

x_i—production of item i
y_i^s—sale for item i in scenario s
z_{ij}^s—substitution sale of item i, satisfying excess demand of item j in scenario s
zt_i^s—substitution sale of item i, satisfying excess demand from all js in scenario s
w_i^s—salvage quantity for item i in scenario s

Parameters:

d_i^s—demand for item i in scenario s
p^s—probability of scenario s
v_i—selling price for item i
c_i—purchasing cost for item i
g_i—salvage value for item i
$\alpha_{ij} \in [0,1]$—substitutability probability; the probability that item j can be replaced by item i

$$\max \quad \sum_{s \in S} p^s \sum_{i \in I} (-c_i x_i + v_i y_i^s + v_i zt_i^s + g_i w_i^s)$$

such that
$$
\begin{cases}
y_i^s + \sum_{j \in I; j \neq i} z_{ji}^s \leq d_i^s & \forall i \in I; s \in S; \\
z_{ij}^s \leq \alpha_{ij}(d_j^s - y_j^s), & \forall i, j \in I, i \neq j; s \in S; \\
zt_i^s = \sum_{j \in I, j \neq i} z_{ij}^s, & \forall i \in I; s \in S; \\
w_i^s = x_i - (y_i^s + zt_i^s), & \forall i \in I; s \in S; \\
x_i \geq 0 & \forall i \in I \\
y_i^s, zt_i^s, w_i^s, \geq 0 & \forall i \in I; s \in S; \\
z_{ij}^s \geq 0, & \forall i, j \in I, i \neq j; s \in S.
\end{cases}
$$

$$(6.2)$$

This maximizes the expected assortment profit from ordinary sales, substitution sales, and salvage. The first constraint set states that total sales for item i—coming from the primary demand for i plus all j sales generated by unmet demand for i—are constrained by the total demand for item i. This constraint can be reorganized as

$$\sum_{j \in I, j \neq i} z_{ji}^s \leq d_i^s - y_i^s \qquad \forall i \in I; s \in S \tag{6.3}$$

stating that substitution sales from item i cannot exceed available unsatisfied demand for i. The next constraint set shows the upper bound on the substitution sale of item i for item j, that is, excess demand for item j with given substitutability probability α_{ij}. Note the interpretation here. Substitution is done in terms of averages, which is necessary since we do not model the individual customers. After having decided who gets their primary wishes satisfied, the manufacturer will, on average, observe that the portion α_{ij} of those wanting item j, but not getting it, will accept item i. We are assuming that at this point the manufacturer can, in fact, find out which ones are willing to take item i instead of j. They are then offered item i, provided there is something to offer according to the first constraint. We are, on the other hand, not assuming that the manufacturer knows up front who will accept substitutes since the retailers are unlikely to reveal such information as it would clearly be used against them in the first round of allocation.

The third constraint set gives the overall substitution sale i from all j's. Next follows the salvage quantity: the quantity of item i left after satisfying primary demand and substitution demand from all j. Finally, nonnegativity constraints are given. These constraints imply that the substitution sale of item i is limited to the remaining supply of the item, that is, $zt_i^s \leq x_i - y_i^s$.

6.5 Test Case and Model Implementation

We analyze a real assortment problem with 15 items from a leading brand name sportswear supplier. Here we attempt to avoid simplifications on the uncertainty and dependencies. The variants within the group are distinguished by the trend driver attributes style and color. We study whether misspecifying substitutability, marginal distributions, and correlations has significant effects on the portfolio structure and its profit.

The true situation of this real assortment problem is one of bimodally distributed individual item demands for about half of the portfolio and dependencies expressed by a correlation matrix with heterogeneous values. This case is denoted as bimodal. For further details see [53]. To say something about the effects of misspecifying the uncertainty and dependencies, we define additional situations, incorrectly assuming a unimodality in demand distributions. More precisely, log-normality is used here, with means and variances as in the bimodal distributions. We analyze this under three different dependency patterns: a correlation matrix with all entries equal to zero, a correlation matrix with all entries equal to 0.5, and the true matrix with heterogeneous values. The incorrect unimodal cases are denoted LogN-$c = 0$, LogN-$c = 0.5$, and LogN-$c = $ true.

Profit and production levels are analyzed while varying a homogeneous substitutability in $[0, 0.5]$, in addition to using the true matrix (denoted mix). Average substitutability values over 0.5 are unrealistic across a group of 15 items. Note that the test cases with the true correlation matrix (bimodal and LogN c =true), having both negative and positive values, are not well suited to directly conclude on the effect of correlations. To do this, we in addition will compare the test results of LogN-$c = 0$ and LogN-$c = 0.5$ (Table 6.1).

Table 6.1: Test cases

Subcase	Marginal distributions	Correlation values c
Bimodal	Bimodal	True matrix
LogN-$c = 0$	Unimodal	Assumed zero correlations
LogN $c = 0.5$	Unimodal	Homogeneous values $c = 0.5$
LogN-$c = $ true	Unimodal	True matrix

6.5.1 Test Results

The expected profit and corresponding production levels for varying substitutability, under the four distributional assumptions, are given by Figs. 6.2 and 6.3. The main point here is how different this production planning problem looks in terms of profit and production depending on the assumptions made about demand and substitution.

Although these results confirm previous qualitative findings, earlier findings say nothing about the effect of misspecifying distributions and dependencies. For this, decisions based on incorrect assumptions must be measured relative to the true distributions and dependencies.

6.5.1.1 The Effects of Misspecifying Substitutability

Here we analyze the effects of misspecifying substitutability, replacing the true substitution willingness (mix matrix) by a matrix with its average (0.15) in each element. For all test cases, the expected profit under the average substitutability (as measured within the model) is found to be higher than the expected profit using the mix matrix (Fig. 6.4). The true bimodal case

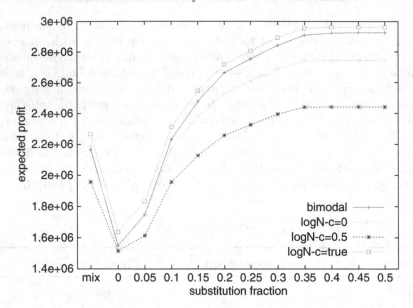

Fig. 6.2: Expected profit versus substitution under the four distributional assumptions

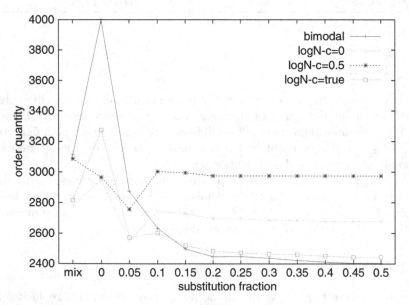

Fig. 6.3: Expected sales versus substitution under the four distributional assumptions

results in almost 15% higher expectation under the average substitutability than under the mix matrix (2,478,901 versus 2,166,684).

However, this does not provide the true picture. If production decisions correspond to the average substitutability, but the true substitution willingness describes the world, decision makers will end up over 30% below their expectation (1,705,694 versus 2,478,901). Low production levels under average substitutability (2,498 versus 3,112 units; Fig. 6.5)—implying low flexibility to adapt to changes when the truth turns out to be the mix substitution matrix—explain the large error in expectations. The substantially lower error (12%) when comparing LogN-$c = 0.5$ with bimodal is due to the strong positive correlation among the items. Substitutability cannot truly be leveraged, as the products mostly face stockout or overproduction simultaneously. The optimal plans suggest almost equally high production levels (2,995 versus 3,087), and hence we observe reduced profit loss. The effects of substitution, and that of misspecifying it, are less significant when the products are strongly positively correlated.

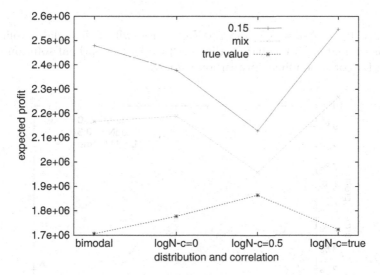

Fig. 6.4: Expected profit under the "mix" substitutability matrix and its average 0.15, evaluated for the four test cases. The *lower curve* shows the true profit when the average substitutability 0.15 is assumed, but the world is described by the bimodal test case and the mix matrix

6.5.1.2 Effects of Misspecifying Distributions and Correlations

Figure 6.6 illustrates the effects of misspecifying distributions by incorrectly assuming unimodality when the world is described by bimodal distributions. We evaluate production decisions obtained from the unimodal cases

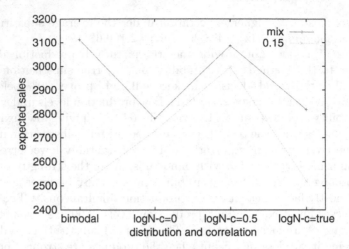

Fig. 6.5: Order quantity for the four test cases, under the assumptions of the mix matrix and its average 0.15

LogN-c = 0, LogN-c = 0.5, and LogN-c = true within bimodal. Profit loss, then, is evaluated by comparing the results with the optimal solution of bimodal, for corresponding substitution values.

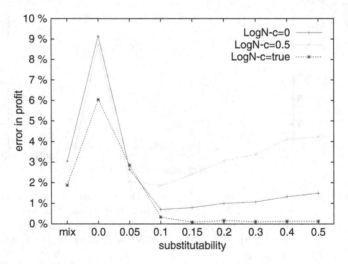

Fig. 6.6: Minimal profit error versus substitution for the unimodal test cases measured within the true bimodal test case

We see, again, that substitution partially compensates for lack of information. When there is no substitution in the group, the error is lowest when at least the true correlation patterns are captured, when LogN-c = true is used. The heterogeneous correlation values, and especially the negative correlations, imply some hedging (see also production decisions for LogN-c =true in Fig. 6.3), contributing to the reduction in error when bimodal actually describes the true world.

The preceding error is studied under maximal uncertainty, assuming that the two states of the world describing the demand uncertainty occur with equal probabilities. In a next step, we measure the negative effects of incorrectly assuming unimodality as information about the items' popularity is revealed. In other words, loss in expected profit is evaluated while increasing the belief about State 1 [probability (State 1) > 0.5], for the unimodal cases LogN-c = true and LogN-c = 0. Production decisions are reoptimized for all cases at hand and for all information levels investigated. The results are summarized by Figs. 6.7 and 6.8. Figure 6.8 gives the profit loss when, in addition to the incorrect unimodal assumption, correlation values are also incorrect (assumed to be zero). Although substitution partially compensates the negative effects of misspecifying distributions and correlations, Fig. 6.7 shows that, even with very accurate information [probability (State 1)= 0.9], and even when the true correlation matrix is used, distributional assumptions are important; the error is up to 16%.

> Resorting to simplifications in distributional assumptions to obtain analytical results can lead to very bad decisions.

This analysis, hence, emphasizes the importance of the underlying distributional assumptions when developing forecasting/planning tools for practical implementation, even when technology and supply chain flexibility allows for continuous information and production updates. To be able to use complicated distributions, we relied on the results of Sect. 4.4.2.

6.5.1.3 The Substitutable Portfolio Structure

The previous sections provided important insights into the substitutable assortment's sensitivity with regard to the values of crucial demand-driver parameters. Here we show the numerical formulation's usefulness when defining the final portfolio structure. Table 6.2 gives the individual item production quantities for the true bimodal case when varying substitutability. For better visualization, production levels under 20 units are eliminated. We observe that the optimal portfolio profit implies trimming some of the products. These products, individually, contribute to the performance and are initially included in the portfolio. However, their substitutability for other products makes them redundant. Observe also the substantial difference in decisions under true substitutability (mix) and when assuming the homogeneous average 0.15 across the group.

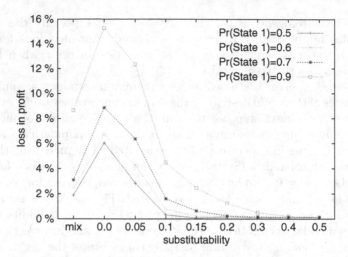

Fig. 6.7: Percentage loss in expected profit under test case LogN-c = true
versus bimodal; evaluated for increasing belief in State 1 for different substitution values

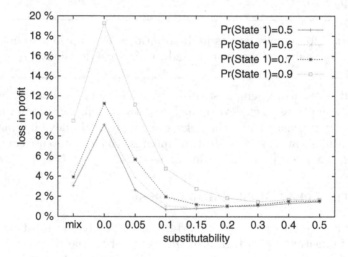

Fig. 6.8: Percentage loss in expected profit under test case LogN-c = 0 versus bimodal; evaluated for increasing belief in State 1 for different
substitution values

Table 6.2: Changes in production quantities for different substitutability values

Subst.	1	2	3	4	5	6	7	8	9	10	11	12	13	14	15	Total
Mix	93	579	486	56	197	84	197	243	194		172	784			21	3,105
0	389	393	393	152	152	152	150	153	153	150	389	347	334	334	358	3,999
0.05	354	384	378	161	160	162	167	163	163	52	402	181	45	43	59	2,874
0.1	220	410	338	102	111	120	299	190	190		179	311			161	2,632
0.15	327	426	424				317	115	123		274	376			115	2,498
0.2	280	512	508				333				314	468				2,435
0.25	155	627	619				249	26			207	561				2,444
0.3	79	732	734				128				100	638				2,411
0.35		833	832				39				32	658				2,393
0.4		839	852									681				2,372
0.45		810	875									699				2,383
0.5		801	831						21			752				2,404

The results in this chapter are potentially important in assortment planning and could not have been achieved by existing analytical formulations. These models have limited potential in even detecting complex effects of misspecifying uncertainty and dependencies.

Our case provides empirical results on how true problem complexity affects the outcome of planning models. In particular, we show the internal uncertainty introduced by commonly accepted and applied simplifications on uncertainty and dependencies among items. In fashion and sports apparel, as well as in other related areas with strong uncertainty about which items become popular, assuming unimodality in demand distributions seriously affects the quality of the planning. The effects of these general underlying assumptions are significant even when it is possible to continuously update information and production decisions. This understanding is potentially important with regard to evaluating appropriate planning softwares available for industrial applications. Underlying assumptions in these tools are largely invisible for practitioners.

6.6 Conclusion

This chapter illustrates the value of using a numerically simple stochastic programming formulation of the multi-item substitutable news vendor problem. Existing analytical formulations are general in their findings; however, solution heuristics and simplifications on the dependencies are frequently applied, and hence it is not at all clear how to read the results.

The stochastic program is simple and handles real problems of substantial size. The CPU times are negligible. As such, the program is potentially useful in practical applications, particularly as fitted to treat the complex nature of demand uncertainty and dependencies observed in quick-response industrial environments such as fashion and sports apparel. The presented approach is an appropriate decision support tool in assortment planning, not just in defining optimal inventory levels, but also suggesting structural changes where appropriate, such as product line trimming. This is clearly important when the negative effects of variety explosion are substantial for suppliers or retailers and when the focus is changing from increasing variety to satisfy heterogeneous customer needs to defining more efficient portfolio planning strategies.

We analyzed how different parameter values affected the assortment planning problem and, hence, the importance of being able to include them in models. In particular, and similarly to financial problems, we observed the impact of dependency patterns in a product portfolio. That said, we pointed out that the error of misspecifying distributions and dependencies is large. The value of substitution is high and compensates, to some extent, for a lack of information, however, only given that the true substitution pattern is captured and quantified by the model. We found up to 30% increase in expected profit when applying our assumed true substitutability matrix versus the common simplification of using average values. The difference stems from structural differences in the first-stage inventory/production levels. We also showed that even when technology and supply chain flexibility allows for continuous information and production updates, the underlying distributional and dependency assumptions used in the planning models are crucial. We believe this observation can be useful in developing decision support tools for industrial use.

Chapter 7

Stochastic Discount Factors

In this chapter, we discuss ways to use information from financial markets to calibrate models for discounting future risks. This type of information is important for modeling the impact of future uncertainty on present decisions. In some areas of activity, there exist well developed financial markets with hordes of traders using the tools and information available to them to decide the present value of future events. This chapter describes a methodology to use market information.

7.1 Financial Market Information

One major question in modeling is where do discount rates come from? Financial market information consists of the prices quoted and volumes traded of financial securities—stocks, bonds, forwards, futures, and options—on public exchanges.

Securities are contracts with standard terms and conditions. For example, a share of a company is a contract that gives the owner the rights to a share of the dividends, and a bond specifies a schedule of coupon payments. Many standardized contracts used by industries are traded on exchanges.

The manufacturer of fashion goods from Chap. 6, for example, will need to ship its goods from where they are manufactured to where they will be sold. Because of the seasonality and the multiple suppliers, it is quite likely that there will be competition for transportation services in these markets. Transportation costs can be quite significant, and the unit prices can be quite variable. However, these services are entirely standard and there exist well-developed markets in shipping contracts.

It is possible to purchase *forward contracts* or *futures contracts* to ship containers of a certain dimension between major hubs like Hong Kong and New York. On the other hand, the manufacturer could choose to wait until the last minute and purchase transportation services on the *spot market*.

A.J. King and S.W. Wallace, *Modeling with Stochastic Programming*,
Springer Series in ORFE, DOI 10.1007/978-0-387-87817-1_7,
© Springer Science+Business Media New York 2012

Of course, relationships must exist between these prices. For example, the Hong Kong to New York price must be related to the Hong Kong to Los Angeles and the Los Angeles to New York prices. If there are any discrepancies, then speculators would enter the market and try to make money from the price differences. One of the key mechanisms by which speculators move in and out of these markets is through options.

7.1.1 A Simple Options Pricing Example

To get started, let us consider a very simple example. Our manufacturer wishes to enter an options contract to purchase a standard contract for transportation services at a future date.

To keep things simple, let us design a simple option that gives the manufacturer the right to ship 100 containers in one year's time at a fixed strike price (a call option) that we suppose to be $8,000. And suppose the manufacturer is experienced enough to know that, historically, there are really only two interesting alternatives for the spot price:

- The spot can either go up to $10,000 (because the ships are full).
- Or it can go down to $7,000 (because the ships are empty).

Now in the first case, where the spot increases in value to $10,000, the manufacturer naturally will exercise the options and save himself $2,000 in shipping costs (we will ignore the effect of interest rates on the amount invested in the option). So the value to him is the $2,000 he has saved. But if the spot decreases in value to $7,000, then the manufacturer will let the options expire, ship at the spot price, and save nothing. Table 7.1 illustrates the example.

Table 7.1: Payout diagram

Period 0		Period 1	
	Scenario 1	Future value	= $10,000
		Option payout =	$2,000
Forward value = $8,000			
Option value = ??			
	Scenario 2	Future value	= $7,000
		Option payout =	$0

How can the manufacturer calculate the value of the option? A trader would answer this question quite simply. She would say that the value of the option is the value of the portfolio that *replicates* the option payouts. Moreover, she would employ a little trick (more on this in a moment) and immediately declare that the value of the option, to the nearest penny, is $666.67!

Before we explain the trick, we will first verify that $666.67 would indeed replicate the option payouts. To simplify matters, we have already supposed that the value of the reference security—the bond—is always $1. Here is how the trader does it.

1. Begin with an amount of cash $666.67.
2. Borrow 4,666.66 units of the reference security (the bond), for a total of $5,333.33 in cash, and purchase 2/3 forward contracts at $8,000 per contract.
3. Liquidate the portfolio to generate the payments:
   - If the spot goes up to $10,000 (scenario 1), then sell the 2/3 forward contract, earning $6,666.67, and use $4,666.66 to pay back the loan. The remaining amount in the portfolio is $2,000.01, which is the required $2,000 option payment (plus a penny rounding error).
   - If the spot goes down to $7,000 (scenario 2), then liquidate the portfolio, as above. The value of the forwards sold exactly equals what you need to pay back the loan, so the amount remaining in the portfolio is $0.

On this very simple sample space, we see that the trader can *replicate* the option payouts provided she has at least $666.67 to start with. If the trader starts out with more than this, then under every possible scenario she will make a profit. This kind of riskless profit making is called *arbitrage*. Market models assume that there are so many traders trading that these kinds of arbitrages are eliminated by competition.

Mathematically, what we need to do is as follows: we must verify that there exists a portfolio $[x_B, x_F]$, where x_B is the number of bonds the trader buys (or borrows) and x_F the number of forwards she purchased (or sold) that satisfies the following constraint system:

$$
\begin{aligned}
x_B + 8{,}000x_F &= 2{,}000/3, \\
x_B + 10{,}000x_F &\geq \quad 2{,}000 \quad \text{(scenario 1)}, \\
x_B + 7{,}000x_F &\geq \qquad 0 \quad \text{(scenario 2)}.
\end{aligned}
\tag{7.1}
$$

The first constraint is the budget constraint for the trader. (We have used exact numbers to make the system easier to solve.) You can verify that a solution to this system is

$$
\begin{aligned}
x_B &= -14{,}000/3, \\
x_F &= \qquad 2/3.
\end{aligned}
$$

A budget of \$666.67, and the capability to borrow a large sum of money from a bank, is what is required for the trader to replicate the option. The option cannot be replicated for less, and if we assume the market in options is free of arbitrage, then the option cannot be worth more. So one can say that the arbitrage-free value of the option is \$666.67.

But what was the trader's mental trick? Is the trader so smart that she can solve a linear program in the blink of an eye?

Well, we said there is a trick in here somewhere. Let us look again at problem (7.1). Notice that if the trader is operating in a competitive, arbitrage-free market, then she can only expect to achieve the *minimal* quantity of starting capital that replicates the option. Let us model this requirement as a linear program with an objective function that consists of this starting capital, so the goal is to use as little cash as possible to satisfy the two scenario constraints. First, express the primal constraint system in matrix form:

$$\begin{bmatrix} 1 & 10{,}000 \\ 1 & 7{,}000 \end{bmatrix} \begin{bmatrix} x_B \\ x_F \end{bmatrix} \geq \begin{bmatrix} 2{,}000 \\ 0 \end{bmatrix}. \tag{7.2}$$

Now let us develop the dual problem. (When an operations research textbook mentions a trick, you will usually find it in the dual!) Assign dual variables q^i to each of these constraints.

By linear programming duality, the dual variables are nonnegative because the primal constraint rows have lower bounds. The constraint matrix is the transpose of the primal matrix, with the right-hand side equal to the primal objective function coefficients. The dual constraint system is

$$\begin{bmatrix} 1 & 1 \\ 10{,}000 & 7{,}000 \end{bmatrix} \begin{bmatrix} q^1 \\ q^2 \end{bmatrix} = \begin{bmatrix} 1 \\ 8{,}000 \end{bmatrix}. \tag{7.3}$$

The dual constraint relationships are of an equality type because the primal variables are unconstrained, and its objective function will be a maximization with objective function coefficients equal to the right-hand side of the primal problem. It follows that the dual problem is

$$\max_{q \geq 0} \quad 2{,}000q^1 + 0q^2$$

$$\text{such that} \quad \begin{cases} q^1 + q^2 & = 1, \\ 10{,}000q^1 + 7{,}000q^2 = 8{,}000, \\ q^1, q^2 & \geq 0. \end{cases} \tag{7.4}$$

Now let us interpret it. The first and last constraints in the dual problem show that the q's are the weights of a probability distribution because they are nonnegative and sum to one. The second condition is a kind of integration

constraint that says that the expected value of the future spot prices using
the probability distribution q equals the price today. The technical term for
a distribution that satisfies a condition like this is *martingale*.

So this is the trader's trick! She knows this basic duality (it is quite general)
and immediately sets about calculating the martingale distribution: she finds
probability weights that satisfy

$$10{,}000q^1 + 7{,}000q^2 = 8{,}000.$$

There is *exactly one* answer: $[\frac{1}{3}, \frac{2}{3}]$. For this answer the value of the dual
objective is

$$2{,}000\frac{1}{3} + 0\frac{2}{3} = 2{,}000/3.$$

Since there is only one answer, we can conclude that this is in fact the optimal
value of the dual problem!

This is why the trader can come up with the price so quickly. She knows
that in this system there is only one dual solution and that it is a martingale
distribution. All she has to do is find it and compute the integral of the option
payouts to determine the optimal value.

This simple options pricing trick holds a number of important points to
keep in mind as the material gets more complicated.

7.1.1.1 Duality and the Martingale Distribution

First (and perhaps most) of all, options pricing is a great example of the
importance of *duality* in modeling. The dual form of the options pricing problem is much easier to solve than the primal, and moreover its solution can
be interpreted as a martingale distribution. In many probability settings, the
martingale condition turns out to have strong implications. For instance, when
asset returns are driven by exponential Brownian motions, as is the case in
many applications of financial mathematics, the martingale distribution turns
out to be unique. The major challenge in these settings is how to calculate the
payouts and perform the integrations for the hypothesized stochastic process.
But the general outline of the solution is the same as in our simple linear
programming setup.

7.1.1.2 Calibration

Secondly, the example shows that prices of derivative securities in a market
model are strongly related to the properties of the asset price change distribution. The martingale distribution in the simple example depends on the
arrangement of future states of price changes. Of course, we chose these particular values to make a nice story. But how do traders price options in a real
market? The detailed answer is complicated, of course, because in real financial markets, the details matter. The timing of dividend or interest payments,

the number of trading days until expiration, carrying costs, opportunity costs, the modeling of option exercises, and the actual asset price processes all have important bearing on valuations. However, the basic answer is always the same. Options are priced in real markets by first finding the martingale distribution that best fits the observed market data. This process of inference is called *calibration*, which we will discuss in subsequent sections.

7.1.2 Stochastic Discount Factors

The pattern used by traders to price options relies on the duality between volumes of trades needed to hedge future outflows and the price distribution to apply to evaluate future outflows. The general framework of duality links the technology of trading (e.g., how efficient is it?) with structural aspects of the price distribution (e.g., is it a martingale?). In our simple example we saw that a trading program with zero transaction costs had dual solutions that were martingale distributions. Additional modeling conditions, such as transaction costs, limits on borrowing, and so forth, create a trading program with dual solutions that may no longer have the martingale property.

These results are well known and fundamental. These dual prices were first observed in their generality by Harrison and Pliska [19] in their examination of the theoretical underpinnings of market equilibrium in contingent claims. The dual variables attached to the replication conditions for the various states of the market are often called *Arrow–Debreu prices* in the economics literature. In the finance literature, the operation of finding the martingale measure and solving the integration problem is called *risk-neutral pricing* because the option is completely replicated and there is no risk. If there is risk, then it would be more appropriate to apply some form of expected utility—such as the Markowitz mean variance or the Kelly growth criterion—to the gap between what can be achieved by trading and the actual option payouts.

We prefer to use a different term that has also appeared in the literature. Our reasons are that, on the one hand, the names Arrow and Debreu would imply an economywide pricing infrastructure, and on the other hand, we do not suppose that risk vanishes in our formulations. Rather, we prefer to use the term *stochastic discount factors* (SDFs) because we intend to apply them in contexts where discounting is required and uncertainty is important.

7.1.3 Generalizing the Options Pricing Model

Let, as usual, $s \in S$ denote the scenarios, possibly developed as in Chap. 4. We let J denote the set of securities, with $j = 0$ the reference security, typically a bond. Further, the vector c denotes the present value of all the securities, while the vector c^s denotes the values in scenario s. The required value payoff from the portfolio in scenario s is f^s.

Traders price options by *replication*, in other words, they try to find the cheapest trading strategy that would replicate those option payouts. A two-stage model of option replication can be modeled as a linear program:

$$F := \min\{cx | c^s x \geq f^s, \quad s \in S\}. \tag{7.5}$$

The way this system works, from the trader's point of view, is this.

1. Begin with an amount of cash F.
2. Allocate it at today's prices, c, into a portfolio x of positions in each underlying security. Positions can be positive (long) or negative (short—a loan, in other words).
3. At each node of the scenario tree liquidate the portfolio, earning $c^s x$ to generate at least the required payment f^s.

The optimal value F is the smallest amount that can replicate all payments without ever falling short. The analysis of this simple system will enable us to derive powerful conclusions about price relationships in a market model.

SDFs are used in market models to compute the fair value of uncertain cash flows that are dependent on market observables. In this section we develop a methodology to calibrate the distribution of an SDF from information contained in market prices of related securities.

To apply the SDF methodology, we return to the two-stage replication problem (7.5) and write down its dual formulation:

$$\max_{y \geq 0} \quad \sum_{s \in S} f^s y^s$$

$$\text{such that} \quad \begin{cases} \sum_{s \in S} c_j^s y^s = c_j & j \in J \\ y^s \geq 0 & s \in S. \end{cases} \tag{7.6}$$

So the dual solution $y = (y^1, \ldots, y^{|S|})$ is nonnegative and has the same dimensionality as the sample space. We call it an SDF for reasons that will become clearer in the following paragraph.

Readers familiar with the financial treatment of options pricing will recall that it is possible to recover a probability distribution from the SDF by discounting. It is not hard to see that the weights q^s given by

$$q^s := y^s \frac{c_0^s}{c_0}$$

are the weights of a probability distribution since this transforms the constraint for the reference security $j = 0$ in (7.6) into $\sum_{s \in S} q^s = 1$. We can think of an SDF in this setting as the product of two terms

$$y^s := q^s \frac{c_0}{c_0^s},$$

where q^s is a probability and c_0/c_0^s is a discount factor. The name *stochastic discount factor* seems to fit this situation well; moreover, it can also be applied to models where the dual variable is not a *martingale distribution*. We prefer to use the term SDF for all these settings.

7.1.4 Calibration of a Stochastic Discount Factor

If we can observe the market prices of related securities, then a simple way to set the weights is to express the observed prices as constraints on the SDF itself. We can write down constraints that require the integral of the option payouts to equal (or almost equal) the observed option's price, as in [36].

Suppose the market has a set I of listed options with observed bid prices, G_i^b, observed ask prices, g_i^a, and future payouts, g_i^s for $i \in I$, $s \in S$. The unknown SDF for problem (7.6) should satisfy, in addition, the following "calibration" inequalities:

$$
\sum_{s \in S} y^s g_i^s \le g_i^a, \quad i \in I,
$$
$$
\sum_{s \in S} y^s g_i^s \ge g_i^b \Leftrightarrow - \sum_{s \in S} y^s g_i^s \le -g_i^b, \quad i \in I. \tag{7.7}
$$

These constraints are to be added to the constraint system of the dual problem (7.6). Each of these added constraint rows will correspond to new primal variables in the trader's replication problem: $\xi^a \ge 0$, interpreted as long positions in the listed option, and $\xi^b \ge 0$, interpreted as short positions in the listed options. Adding these inequalities results in a new primal problem. The primal objective is formed by taking the product of the dual right-hand side with the new primal variables, resulting in the following calibrated primal problem:

$$
\min_{x, \xi^a, \xi^b} \quad cx + (g^a \cdot \xi^a - g^b \cdot \xi^b)
$$
$$
\text{such that} \quad \begin{cases} c^s x + g^s \cdot (\xi^a - \xi^b) \ge f^s & s \in S, \\ \xi_i^a, \xi_i^b \ge 0. \end{cases} \tag{7.8}
$$

This problem corresponds to allowing the trader to take positions in the market-traded options to hedge the payouts f^s. You may have noted that the dual constraints (7.7) are "hard." What if these constraints cannot be satisfied?

7.1.4.1 Liquidity-Weighted Calibration

It may surprise you—with all of the emphasis on arbitrage and efficient markets and equilibrium—to learn that when actual market data are entered into this problem, the solution (almost always) tries to take an unbounded position in one of the listed options. So, in fact, there is arbitrage in financial markets. How can this be?

The reason has to do with trading volumes, or market liquidity. Sellers and buyers of listed options always indicate how many option contracts they are willing to buy or sell at the advertised price. These offered volumes can be extremely low for options that are lightly traded. If anyone shows interest in buying or selling at the advertised price, then the trader will quickly pay attention. Prices for additional quantities could change substantially.

Thus, it is natural to constrain the number of options available to the model to be less than the total volume of the option traded thus far in the session.

$$\xi_i^a, \xi_i^b \le v_i, \quad i \in I. \tag{7.9}$$

With these constraints, the solution is bounded and quite reasonable results are observed in practice. These constraints are added to the primal problem, so let us see what happens to the dual. The dual will have an additional $2I$ nonnegative variables corresponding to the liquidity bounds (7.9). Let us call these variables λ_i^a and λ_i^b for $i \in I$. Glancing back at (7.7) we can verify, through linear programming duality, that these volume bounds translate into volume-weighted penalty terms for violating the calibration inequalities, expressing the natural observation that the greater the volume traded, the more reliable are the price quotes.

$$\max_{y \ge 0} \quad \sum_{s \in S} f^s y^s - v \cdot [\lambda^a + \lambda^b]$$

$$\text{such that} \quad \begin{cases} \sum_{s \in S} c_j^s y^s = c_j, & j \in J, \\[2mm] \sum_{s \in S} g_i^s y^s - \lambda_i^a \le g_i^a, & i \in I, \\[2mm] \sum_{s \in S} g_i^s y^s + \lambda_i^b \ge g_i^b, & i \in I, \\[2mm] y^s \ge 0, & s \in S, \\[2mm] \lambda_i^a, \lambda_i^b \ge 0 & i \in I. \end{cases} \tag{7.10}$$

In summary, calibration to externally observed option prices is a way of using the information in the market equilibrium to place constraints on the SDF. Calibrated discount factors can be extended by smoothing or some other statistical techniques for use as an approximate discounting operator. In the case of highly liquid markets, the SDF can be quite accurately determined from the observed market prices.

In real situations, there are two issues that must be confronted. First, the manufacturer's actual needs may not be precisely covered by the contracts being offered in the market. For example, the manufacturer may need to ship on September 1, but the contracts specify August 15 and September 15. Or the contracts may be for quantities that are different than the standard contracts. The stochastic programming approach outlined above is easily applied to these kinds of problems. A second, and more challenging, concern is that the standardized contracts may be correlated with the uncertainty but not identical with it.

7.2 Application to the Classical NewsVendor Problem

In this section, we consider how the framework for news vendor models can be extended to address uncertainty from both market and nonmarket sources. Our approach here can be viewed as an extension of Birge [6]. We utilize the model formulation of Gaur and Seshadri [17] as a starting point for this discussion.

We define the following problem variables and parameters:

$$
\begin{aligned}
I &: \quad \text{initial inventory to be ordered,} \\
D &: \quad \text{future demand,} \\
p &: \quad \text{unit selling price,} \\
c &: \quad \text{unit cost,} \\
s &: \quad \text{salvage price,} \\
r &: \quad \text{risk-free interest rate.}
\end{aligned}
\tag{7.11}
$$

The news vendor has a time 0 cash flow that consists of the cost paid for the initial inventory of newspapers

$$
F_0(I) = -cI \tag{7.12}
$$

and a time T cash flow that consists of the income from the sold newspapers plus a salvage term for the unsold inventory:

$$
F_T(I) = p \min[D, I] + s \max[0, I - D]. \tag{7.13}
$$

At all other intervening times, $t \in (0, T)$, there are no news vendor cash flows, so we set $F_t(I) := 0$.

We suppose that a simple statistical model relates future demand D to the price at time T of a certain market-traded security, S:

$$
D = bS_T + \epsilon, \tag{7.14}
$$

where ϵ is a random noise term that is independent of S_T. For example, the market-traded security could be a stock-market index, which the news vendor could relate to sales: high growth of the index would indicate high sales, and poor performance could be indicative of poor sales.

We suppose that $\mathcal{M}_t$ describes the observable states for market prices S_t, and $\mathcal{N}_t$ describes the observable states for the nonmarket sources ϵ. Such a model (7.14) could be constructed by regression of demand data against the security prices. Substituting for D in (7.13) and collecting the terms yields

$$F_T(I) = pI - (p-s)(I-\epsilon) + (p-s)b\big(S_T - \max[0, S_T - (I-\epsilon)/b]\big). \quad (7.15)$$

This equation describes the news vendor's income in terms of $(p-s)b$ units of a long position in security S and $(p-s)b$ units of a short position in a "call option" with strike $(I-\epsilon)/b$. Performing some algebra we can obtain a concise expression for the news vendor's income:

$$F_T(I) = pI - (p-s)b\big[(I-\epsilon)/b - S_T + \max[0, S_T - (I-\epsilon)/b]\big] \quad (7.16)$$

$$F_T(I) = pI - (p-s)b\max[0, (I-\epsilon)/b - S_T]. \quad (7.17)$$

This shows that the news vendor time T cash flow equals the nominal income pI from selling all the inventory minus $(p-s)b$ units with cash flows that resemble the payoffs of "put options" with strike price $(I-\epsilon)/b$. This suggests that if the news vendor wanted to hedge the uncertainty in his business, he should buy put options in the market-traded security.

Put options are well-known vehicles for insuring portfolios. So what we have shown here is that the news vendor, the manufacturer desiring transportation services, or really any business that purchases inventory in advance of sales will be in the market for *insurance* in the form of put options.

Now let us suppose that the news vendor wants to sell his business. How much would someone pay for it? A buyer looks at the future cash flows and tries to determine the *most valuable* portfolio of assets he could buy in the market today that will be paid for by the cash flows from the business. (This is the reverse of the options pricing problem in which the hedger determines the minimum cost portfolio whose cash flows will exceed the cash flows of the option.) The idea here is that the buyer's future cash flows will pay for the initial asset purchase. One can express the news vendor "valuation" problem as follows:

$$\max_{I,\theta,\xi} \quad S_0\theta_0 + (C_a\xi_0^a - C_b\xi_0^b)$$

$$\text{such that} \quad \begin{cases} S_t(\theta_t - \theta_{t-1}) + (C_t\xi_0^a - C_t\xi_0^b) = F_t(I), \\ S_T\theta_T \geq 0, \\ \xi_0^a, \xi_0^b \geq 0, \end{cases} \quad (7.18)$$

where I^* denotes the optimal initial inventory. The optimal value V^* is the upper limit of what anyone would pay for the business.

7.2.1 Calibration of Real Options Models

This approach can also be applied to any real options modeling problem. The basic ideas are as follows:

1. Generate a scenario tree for the underlying process, for example, oil spot prices.
2. Identify some market-traded options in the underlying and model their payouts on the scenario tree.
3. Develop an investment model that allows trading in the underlying, in a bond, and allows the taking of buy-and-hold positions in the options.
4. Apply reasonable volume constraints on the options.
5. Model the cash flows of the real options problem as the underlying option to be hedged.
6. Solve for the minimum cash needed to payout the cash flows. This is the sell price.
7. Solve for the maximum cash that could be obtained from receiving the cash flows. This is the buy price.

If the sell and buy prices are close in value, then you may conclude that the model is giving you good information about the option value. If they are wide apart, then you may wish to consider incorporating a utility function into the problem. For details on this approach, see King [35] and King et al. [37].

7.3 Summary Discussion

In this section, we examine the options pricing model in the spirit of this book. After all, this is a model that claims to offer a methodology to manage uncertainty. Does it really work as advertised?

Of course, discounting is important. One can hardly plan for the future without considering discounting. Ever since the development of options pricing it has been known that the pricing of options is a natural model in which to introduce SDFs as a mechanism for accounting for the impact of future events in the consideration of the costs and benefits of current actions. But let us in any case give the model a more detailed examination.

First, what is the uncertainty? In the simple model of Sect. 7.1.1, the uncertainty space is the shipping costs:

- Forward value of spot price $8,000.
- Spot can either go up to $10,000 (because the ships are full).
- Or it can go down to $7,000 (because the ships are empty).

over which we derive the probability weights of the martingale measure:

$$\max \quad 2,000q^1 + 0q^2$$

$$\text{such that} \quad \begin{cases} q^1 + q^2 & = 1, \\ 10,000q^1 + 7,000q^2 = 8,000, \\ q^1, q^2 & \geq 0, \end{cases}$$

conclude that the optimal value is \$666.67, and derive the optimal hedge: borrow \$4,666.67 and buy 2/3 of a forward contract for \$5,333.33.

Now let us look at the end stage. Denote by ξ the spot price for shipping. Here is what happens:

- The cash flow generated by the option is $\max\{0, \xi - 8,000\}$.
- The cash flow generated by the portfolio is $\frac{2}{3}\xi - 4,666.67$.

Do you see that the portfolio cash flow is linear? We have graphed the two curves in Fig. 7.1.

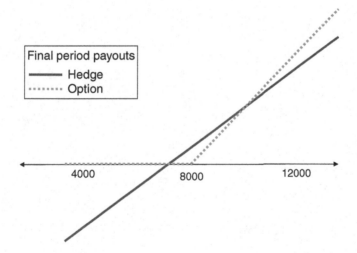

Fig. 7.1: Graph of option cash flow versus hedging portfolio cash flow

Can you see that the linear portfolio cash flow just touches the graph of the option payout at the two points corresponding to \$7,000 and \$10,000? This is the result of insisting on a hedging policy that matches cash flows at these two outcomes. But for values below \$7,000 and above \$10,000 the value of the portfolio lies below the value of the option. Our quick-minded trader did not use a model that took these possibilities into consideration!

The calibration model, on the other hand, can potentially do a better job of matching the option cash flows. For one thing, the optimization is carried out over a wider range of possible outcomes for ξ. For another, the calibration model allows the trader to take positions in listed options. These listed options have nonlinear graphs and so offer a greater capability to match the nonlinear shape of the option cash flows.

Of course, the calibration model is better precisely because some other traders have done the hard work of posting quotes for the listed options. The calibration model specifies a martingale measure using the *already known* prices of the listed options. Our quick-minded trader is *estimating* the martingale measure from a simple two-point approximation. The calibration model is making use of a great deal more information than was available to our quick-minded trader.

Chapter 8

Long Lead Time Production: With Aliza Heching

In the long run we are all dead
– *John Maynard Keynes*

We consider a problem faced by a supplier of custom products that have long production lead times. The problem is inherently multistage with a large number of stages. The majority of these products have no salvage value, so the supplier is exposed to significant risk of excess production. Moreover, customer forecasts will likely err on the upside because the option to purchase has value. This chapter describes a counterbalancing mechanism for the supplier to obtain some compensation for part of the inventory risk.

From a modeling perspective, the most challenging part here is to combine in a meaningful way stages and time periods and then fit in the long lead times. Some of the details will get very involved, but try to follow us on the main modeling questions.

8.1 Supplier-Managed Inventory

A supplier-managed inventory (SMI) program is a way to compensate a manufacturer for inventory risks. The basic steps of the program are as follows:

1. Customers provide the supplier with a demand forecast schedule.
2. The supplier responds with production forecast schedules.
3. The SMI program determines what part of the production forecast falls within the production lead time. This is a *commitment* since it cannot be altered.
4. Periodically—for example, at the end of each quarter—the supplier compares the commitments with the actual customer orders.
5. When orders are lower than commitments, the supplier has the right to ship part of this *underorder quantity* to the customer.

A.J. King and S.W. Wallace, *Modeling with Stochastic Programming*,
Springer Series in ORFE, DOI 10.1007/978-0-387-87817-1_8,
© Springer Science+Business Media New York 2012

The SMI program reduces the incentive for customers to overforecast their demand. Customers have a natural incentive to overforecast because the access to production facilities gives them an option to increase their own production while also denying their competitors access to a scarce production resource.

8.2 Supplier-Managed Inventory: Time Stages

Our goal is to develop a model of the SMI program that indicates the dependence of the supplier's decisions on the information flow. We also illustrate the need for flexibility in modeling decision stages.

8.2.1 Modeling Time Stages

One of the most important details in stochastic modeling is to sort out the stages. Stages define the boundaries of time intervals. The conceit of stochastic programming is that actions and events whose time of occurrence or observation falls *within* a given time interval are actually accounted for at the *end* of that time interval.

Let us think a bit more about why this is so. In the SMI program, the supplier will be collecting information about customer orders and will be making production decisions. At the review point, the supplier closes the books and makes the underorder decision. We imagine the action of closing the books as happening *at the same time* as the history of orders and production since the last review point became known since we make the decision with this information in mind. Technically, this means that time intervals between review points are "open" on the left and "closed" on the right and that events and actions that occur within time intervals are labeled with the right-hand endpoint. Such considerations are only important in models handling information uncertainty. In deterministic models, this issue is irrelevant.

As we have emphasized throughout this book, the first issue of substance when dealing with uncertainty in decision problems is to decide how "clock time" maps to "decision time." Production decisions are made at regular clock-time intervals since every production facility (or, for that matter, any collection of coordinated processes) needs a reference clock to synchronize the component tasks. So from a "clock-time" perspective we say that a certain product takes a certain number of clock-time periods (e.g., minutes, days, months) to be produced, but we need a procedure to map these clock lead times into decision lead times.

The SMI program declares a review-time process that is likely to consist of many clock-time intervals. A natural incentive arises from the *complexity of solving* the problem to reduce the number of time intervals, of course. For this reason alone we will want to focus on review times since these are the least frequent time points that are relevant for the problem. However, easing the burden of solving the stochastic program is a very natural and important consideration.

We need to identify natural features of the modeling that justify the choice of stages. The most natural feature to consider is *when we learn something.*

In stochastic programming, we identify *stages* with *learning.*

At review times the supplier must study the relationship between customer forecast behavior and actual customer orders because this is the supplier's opportunity in the SMI program to counter the overforecasting incentive. Hence the intervals between review times are the natural candidates for decision stages.

Of course, as the review time approaches, the supplier will be able to form hypotheses about the likely underorder quantity at review time. Perhaps it is sensible to subdivide the review intervals to implement decision stages that react to the increasing reliability of the underorder estimates. In natural systems, like processes inside living cells, such anticipation makes sense.

In supply chain models, the supplier cannot control the customer's order behavior or develop perfect visibility into the processes that influence customer order behavior. Customers could decide to place a huge order just before the review period. Such behavior may even make sense, from the customer's perspective, because the customers of our customer may in turn have their own end-of-period incentives. For example, it is well known that budget cycles drive institutional orders. The "use it or lose it" principle drives institutional members to rush to use up their budget allocation before the end of their fiscal year. (Just think about the importance of Christmas for retail sales in countries with a strong Christian tradition!) And of course, a large end-of-period order may just be irrational. The point is that we cannot control it.

The SMI program is a program that enables a supplier to gain more reliable information about customer demand. Of course, it makes sense for supplier and customer to coordinate the review process timing to account for the natural order cycles in their respective businesses.

For concreteness, let us index clock time by $t = 0, 1, \ldots, T$. Time intervals are open on the left, closed on the right, and labeled with the right endpoint, so that events falling in the interval $(t-1, t]$ are said to occur during period t. Production lead time L_i, for now, will denote the number of clock-time periods needed to manufacture product i.

The review process takes place at time steps that are multiples of the clock time. Let us say that $\eta = 1, 2, \ldots$ indexes review times, and t_η is the time stamp for the review time η. Thus at time t_1 the SMI program will review the production completed before t_1, at time t_2 review production completed since t_1, and so forth.

Some gray areas need to be examined here. For example, how do we handle a product that was committed before the review period ends but arrives in the following period? There will always be time-boundary issues like this, and the problem formulation process needs to address special ways to handle them. Such discussions often lead to the development of a deep insight into the decision process. In the SMI model, we can handle this issue by allowing the customer to order a product that is committed to manufacture but has not yet arrived in inventory—but of course this introduces a more complex inventory update equation, as we will see below.

8.3 Modeling the SMI Problem

The actions that occur at every clock-time stage are as follows:

- Customer provides demand forecasts for multiple future clock-time periods.
- Supplier indicates what product is now committed to production, by period.
- Customer places orders.
- Supplier makes production decisions.

At review periods, in addition, there are the following actions:

- Supplier reviews orders and commitments and determines underorder quantities.
- Supplier ships orders, including a proportion of the underorder.

8.3.1 First- and Last-Stage Model

The first stage is always special, and the flexibility to model first-stage considerations is one of the great strengths of the stochastic programming methodology. The first decision can occur at any clock-time point, and because time starts at 0, we say the first stage is $t = 0$. The information observable at the first decision stage is

- Product in inventory.
- Product commitment and product orders since last review period.
- Customer order and supplier production history since the beginning of time.

The second decision stage corresponds to the interval $(0, t_1]$, where t_1 is the first review time following time 0, or "today." The third decision stage will be $(t_1, t_2]$, where t_2 is the second review time, and so forth.

The final consideration is how to handle the terminal stage. Stochastic programming cannot rely on "long-run" limiting arguments, which discount the horizon into irrelevance—in the long run we are all dead, as the great economist pointed out. So we need to decide what the horizon should be and account for the real choices that exist there. The horizon should be beyond the first review time, but how far beyond? Perhaps it makes sense to select a horizon that is the longest lead time for the current planning cycle? Many arrangements are possible here.

To make the development of the model concrete, we will implement the SMI program as a three-stage problem. Stage 1 is "today," or time 0; stage 2 includes the time until the first review point $(0, t_1]$; and stage 3 is $(t_1, t_2]$, where t_2 is the second review point following today.

8.3.2 Demand Forecasts and Supply Commitments

The managed inventory model begins with an exchange of clock-time demand forecasts and supply commitments between customer and supplier. Let the horizon for the forecasts be τ. Denote the customer's forecast issued at clock time t for product i demand during periods $t + 1, \ldots, t + \tau$ by

$$\mathbf{F}_t^i = \left\{ F_t^{i,1}, \ldots, F_t^{i,\tau} \right\}. \tag{8.1}$$

The supplier reviews the forecasts $\mathbf{F}_t^i$ and responds with the volume of product projected to be available for delivery during each future period $t+1, \ldots, t+\tau$. Denote the supplier response by

$$\mathbf{C}_t^i = \left\{ C_t^{i,1}, \ldots, C_t^{i,\tau} \right\}. \tag{8.2}$$

The rules of the managed inventory contract dictate that those forecast periods that fall within the production lead time are *commitments* of the supplier and the customer.

At stage $t = 0$, which in our modeling framework is "today," the commitments are the responses $C_0^{i,1}, \ldots, C_0^{i,L-1}$ because these are already in production. The response $C_0^{i,L}$ is the new commitment.

8.3.3 Production and Inventory

The supplier plans the release of new production and manages the inventory on a clock-time frequency. The release of product into production will create inventory but, due to manufacturing lead times, will not be available until L periods have passed. Most inventory models will incorporate customer deliveries, but in the SMI program, actual customer orders are relevant only at review periods.

Q_t^i = quantity of product i released to manufacturing during period t
X_t^i = on-hand inventory of product i at the end of period t
D_t^i = product i delivered to customer during period t

The supplier makes the production decision at t, but the actual quantities will be released to manufacturing in the interval $(t, t + 1]$. They will arrive at inventory during the interval $(t + L, t + L + 1]$. Stated another way, the production arriving at inventory during interval $(t - 1, t]$ was released to manufacturing L periods before, or $(t - L - 1, t - L]$. Thus, the basic inventory update equation is given by

$$X_t^i = X_{t-1}^i + Q_{t-L}^i - D_t^i. \tag{8.3}$$

The supplier commitments require that product available for delivery be greater than the commitment, namely,

$$X_t^i \geq C_0^{i,t}. \tag{8.4}$$

8.4 Capacity Model

The manufacture of new products consumes the services of labor and equipment. The availability of these resources is constrained. Let us say that a unit of product i at manufacturing stage $k = 1, \ldots, L$ consumes a vector of resource capacities α_{ik}. In this section, we are not concerned with the details of modeling the allocation of resources, so we simply assume the existence of an exogenously given resource capacity schedule A_t. The capacity constraint is then

$$\sum_{k=1}^{L} \sum_{i \in I} \alpha_{i,k} Q_{t-k}^i \leq A_t. \tag{8.5}$$

This is a hard constraint, which should concern you; after all, we have made a big point of using soft constraints. Can you work out a version of this problem that has soft constraints for capacity?

8.4.1 Orders and Review Periods

Customer orders are under the control of the customer and observed by the supplier. Accumulated orders since the last review period are denoted Y_η^i. The accumulated deliveries cannot be greater than the accumulated orders, hence

$$\sum_{t \in (t_{\eta-1} - t_\eta]} D_t^i \leq Y_\eta^i. \tag{8.6}$$

At the review point, the supplier compares the total orders with the commitments and calculates the underorder, namely,

$$U_\eta^{i+} - U_\eta^{i-} = Y_\eta^i - \sum_{t \in (t_{\eta-1}, t_\eta]} C_0^{i,t}, \tag{8.7}$$

where the quantities U_η^{i-} and U_η^{i+} are nonnegative. The underorder is the quantity that picks up the negative amount. According to SMI rules, this amount is available in inventory and a proportion γ_i can be delivered immediately. Hence we need to update the inventory equation in the first period following the review period with the underorder quantity:

$$X_{t_\eta}^i = X_{t_\eta-1}^i + Q_{t_\eta-L}^i - D_{t_\eta}^i + 1_{t=t_\eta}\gamma_i U_\eta^{i-}, \tag{8.8}$$

where the expression $1_{l=l_\eta}$ equals 1 when the subscript condition is satisfied and zero otherwise.

Finally, revenue is generated when product is delivered:

$$R_\eta^i = \rho_i \left[\sum_{t\in(t_{\eta-1},t_\eta]} D_t^i + \gamma_i U_\eta^{i-} \right], \tag{8.9}$$

and we model this as a quantity to be maximized by the supplier.

At this point, poor reader, you do have our sympathies. There is certainly a great deal to keep in mind in this complicated model. Perhaps the following questions may help to direct your attention:

1. Can you verify that our model is actually "implementable"? Do prior time quantities depend directly on future time quantities without provision for recourse? To tell the truth, the authors worried a great deal about this aspect of the problem formulation. Did we get it right?
2. If we presume that the supplier maximizes revenue, do the delivery variables get set to the right quantities?

8.4.2 The Model

We arbitrarily elected to model a problem with three time stages: today, the first review stage, and the second review stage. The order of the arrival of information, the decisions to be made, and the consequences of these decisions are described in Fig. 8.1. It is useful to consult this figure while reading the rest of this section. Note that customer orders and demand forecasts are revealed before the supplier makes supply commitments and production decisions.

8.4.3 Objectives

Large manufacturers typically have three objectives in mind: high customer satisfaction, high net revenue, and low inventory charges. Assuming that the first objective is satisfied by successfully implementing the terms and conditions of the SMI program, we now develop an optimization formulation of a three-stage inventory management model that emphasizes the attainment of revenue and management of inventory expense targets.

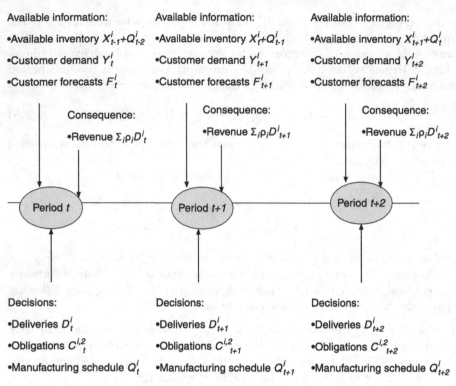

Available information:

•Available inventory $X^i_{t-1}+Q^i_{t-2}$

•Customer demand Y^i_t

•Customer forecasts F^i_t

Consequence:

•Revenue $\Sigma_i\rho_iD^i_t$

Available information:

•Available inventory $X^i_t+Q^i_{t-1}$

•Customer demand Y^i_{t+1}

•Customer forecasts F^i_{t+1}

Consequence:

•Revenue $\Sigma_i\rho_iD^i_{t+1}$

Available information:

•Available inventory $X^i_{t+1}+Q^i_t$

•Customer demand Y^i_{t+2}

•Customer forecasts F^i_{t+2}

Consequence:

•Revenue $\Sigma_i\rho_iD^i_{t+2}$

Period t Period $t+1$ Period $t+2$

Decisions:

•Deliveries D^i_t

•Obligations $C^{i,2}_t$

•Manufacturing schedule Q^i_t

Decisions:

•Deliveries D^i_{t+1}

•Obligations $C^{i,2}_{t+1}$

•Manufacturing schedule Q^i_{t+1}

Decisions:

•Deliveries D^i_{t+2}

•Obligations $C^{i,2}_{t+2}$

•Manufacturing schedule Q^i_{t+2}

Fig. 8.1: Information revealed/decisions over time

Let

R_η = net revenue target in review periods $\eta = 1, 2$

E_η = inventory expense target in periods $\eta = 1, 2$

ρ_i = revenue per unit of item $i \in I$ ordered

χ_i = inventory cost per unit of item $i \in I$ in on-hand inventory

Expense and revenue targets will be enforced through the objective function as "soft constraints" since it may be impossible to satisfy both targets simultaneously. Also, as *targets* they obviously are not absolutely required, and, hence, the target constraints belong among the soft constraints.

> Targets should always be modeled in a soft way by incurring a penalty whenever they are broken.

To that end, we introduce nonnegative variables $Z_{R\eta}^+$ and $Z_{R\eta}^-$ and an equality constraint that defines the variables as the excess and deficiency, respectively, of the revenue with respect to the target:

$$\sum_{i \in I} R_\eta^i - Z_{R\eta}^+ + Z_{R\eta}^- = R_\eta.$$

To penalize solutions that do not achieve revenue targets, we incorporate the penalty term $-\lambda Z_{Rt}^-$, with $\lambda > 0$, into the objective function. Similarly, we introduce nonnegative variables Z_{Et}^+ and Z_{Et}^- and define them as the excess and deficiency, respectively, of the inventory expense relative to its target:

$$\sum_{i \in I} \chi_i X_\eta^i - Z_{E\eta}^+ + Z_{E\eta}^- = E_\eta.$$

A second term added to the objective function, $-\mu Z_{E\eta}^+$, with $\mu > 0$, penalizes solutions with above-target inventory holding costs. Of course, the penalty factors λ and μ can depend on i and η. The supplier may wish additional parametric degrees of freedom to adjust penalties by product or review time—for example, to assess the impact of the inventory charges relative to the production capacities.

Since this style of soft constraints penalizes falling on the wrong side of a target, these are sometimes referred to as "shortfall penalties"; see Sect. 3.4 for a discussion of soft constraints and targets.

8.5 Uncertainty

There are many potential sources of uncertainty. The two major sources are the capacity model (8.5) and the customer order processes, Y_η^i. In this chapter we focus on uncertainty associated with customer orders.

8.5.1 Uncertain Orders

The SMI agreement requires a customer to provide τ period rolling horizon order forecasts. The supplier has different uses for forecasts with different lead times. Longer lead-time demand forecasts are typically used for strategic planning. Shorter lead-time forecasts are used to schedule actual production. Intermediate lead-time forecasts are used for capacity planning; the supplier reviews expected demand and may decide to acquire capacity from outside sources if demand forecasts indicate that demand exceeds production capacity. Thus, the supplier faces multiple types of risk due to the uncertainty associated with the demand forecasts with different lead times. Further, errors in forecasts at different lead times can have very different financial impacts on the supplier.

The model under consideration in this chapter addresses risks of inventory imbalances by modifying production schedules. Therefore, what is of most interest is the short-term forecast uncertainty.

Suppliers depend upon customer forecasts for strategic and tactical planning, but the reasons why customers may report inaccurate order forecasts are quite varied.

8.5.2 Inaccurate Reporting

When customers do not anticipate tight production capacity, they may often treat these forecasts as a courtesy to the supplier. Thus, customers may not invest adequate time in properly estimating expected demand. In particular, where the penalty associated with inaccurate forecasts is not significant, the supplier may see sizeable discrepancies between the forecasted demand that the customer reports and the realized demand.

Anticipation of constrained supply may also lead customers to provide the supplier with inaccurate demand forecasts. When a customer anticipates that production capacity is constrained, the customer may forecast higher-than-expected demand on the assumption that the rationed quantity of product that he receives is closer to his true needs.

Discrepancies between forecasted and realized demand are not always due to intentional actions taken by the customer. For example, if a supplier is manufacturing custom products, forecasted demand for a specific prototype will not materialize into actual demand if changes are made to the prototype. Forecasts for products with mature and successful designs will not suffer from this specific cause for forecast uncertainty.

As another example, customers may be incorporating these products into end products that have a "fashion" component, i.e., where end-consumer interest cannot be established up front. Thus, once the customer begins to sell the end product on the retail market, he may find that end-consumer interest is not as anticipated. For a more direct discussion of the difficulties of "fashion" products, see Chap. 6. Finally, unanticipated changes in global market conditions can impact demand and cause actual demand to diverge from forecasted demand.

8.5.3 A Stochastic Programming Model

A model's stochastic version models uncertainty in demand forecasts. As we indicated above, we wish to simplify the model by incorporating uncertainty only at review periods. There will be different realizations of forecasts and orders depending on what scenario $s \in S$ occurs. The objective function is an integral over the scenarios, weighted by probabilities p_s.

Then the supplier's objective is to solve

$$\max_{(Q,\mathcal{C},\mathcal{D},U)} \sum_{s\in S} p_s \sum_{\eta=1}^{2} \left[\sum_{i\in I} R_\eta^i(s) - \lambda Z_{R\eta}^-(s) - \mu Z_{E\eta}^+(s) \right],$$

$$\begin{aligned}
\sum_{i\in I} R_\eta^i(s) - Z_{R\eta}^+(s) + Z_{R\eta}^-(s) &= R_\eta, \\
\sum_{i\in I} \chi_i X_\eta^i(s) - Z_{E\eta}^+(s) + Z_{E\eta}^-(s) &= E_\eta, \\
\rho_i \sum_{t\in(t_{\eta-1},t_\eta]} D_t^i(s) + \gamma_i U_\eta^{i-}(s) &- R_\eta^i(s), \\
U_\eta^{i+}(s) - U_\eta^{i-}(s) + \sum_{t\in(t_{\eta-1},t_\eta]} C_0^{i,t} &= Y_\eta^i(s),
\end{aligned}$$

such that

$$\begin{aligned}
\sum_{t\in(t_{\eta-1}-t_\eta]} D_t^i(s) &\leq Y_t^i(s), \\
X_t^i(s) - X_{t-1}^i(s) - Q_{t-L}^i(s) + D_t^i(s) + 1_{t=t_\eta}\gamma_i U_\eta^{i-}(s) &= 0, \\
X_t^i(s) - C_0^{i,t} &\geq 0, \\
C_0^{i,t} &\leq F_0^{i,t}, \\
\sum_{k=1}^{L} \sum_{i\in I} \alpha_{i,k} Q_{t-k}^i(s) &\leq A_t, \\
Q, \mathcal{C}, \mathcal{D}, X, U, Z &\geq 0,
\end{aligned}$$

$$(8.10)$$

where all the appropriate initial values are given. (Can you determine what these should be?) This formulation is a three-stage stochastic program with recourse. Decisions in the second stage are made after having observed actual events in the first stage, and so forth.

8.5.4 Real Options Modeling

A stochastic programming formulation requires the supplier to model risk preferences and distributions. The previous formulation used a target penalty approach to deal with the uncertainty in revenues and expenses. But where did the probability measure come from?

We close this chapter with a brief discussion of this issue in the light of "real options modeling" (ROM), as we described it in Chap. 3. The basic idea is to apply *forward-looking information*, such as may be found in financial markets, to evaluate the risk of operational decisions using a *stochastic discount factor* (SDF) as presented in Chap. 7. The key step in the ROM approach is to develop a state space from which we can obtain calibrations for the SDFs, as discussed in Sect. 7.1.4.

The main source of uncertainty we consider is the uncertainty in customer orders Y_η. We require sources of information that can calibrate SDFs on this (or, perhaps, a related) state space. Some possible sources of information include:

- Forecasts for customer orders.
- Internal opinions on forecast errors.
- Supplier-company options, given some model in which supplier-company stock values Or payments are impacted by forecast variations.
- Customer-company options, given some model in which customer-company stock values are affected by forecast variations.

The supplier's internal opinions on forecast errors are probably the most valuable source of calibration. In Chap. 3 we suggested a form of internal market where employees could buy or sell option contracts on future customer revenues. These option contracts can be modeled as virtual "hedges" for the revenue and expense cash flows. Employee opinion generated by the creation of an internal exchange of option contracts would probably be the best source of opinion. Of course, the validity of opinions can be weighted by the volume of the bets entered into the exchange, as discussed in Sect. 7.1.4.1.

One way to obtain option information in the case of large corporate entities is to look at options that are traded on open exchanges. It would be logical to think that robust order growth from a customer should be reflected in the stock prices.

However, there are difficulties with this source of information. First, multiple factors drive stock prices, and the line of business modeled by SMI may only be one of these. Second, the trading public may not have sufficient visibility on the SMI business to form opinions regarding its likely impact on stock prices. The authors themselves have observed examples where customer orders shot up but customer-company stock first declined followed by a sharp increase more than a year later.

References

[1] Laurent El Ghaoui Aharon Ben-Tal and Arkadi Nemirovski. *Robust optimization.* Princeton Series in Applied Mathematics. Princeton University Press, 2009.

[2] M. Ball, C. Barnhart, G. Nemhauser, and A. Odoni. Air transportation: Irregular operations and control. In C. Barnhart and G. Laporte, editors, *Transportation*, number 14 in Handbooks in Operations Research and Management Science, chapter 1, pages 1–67. Elsevier, 2007.

[3] Güzin Bayraksan and David P. Morton. Assessing solution quality in stochastic programs. *Mathematical Programming*, 108(2–3):495–514, sep 2006. doi: 10.1007/s10107-006-0720-x.

[4] Güzin Bayraksan and David P. Morton. A sequential sampling procedure for stochastic programming. *Operations Research*, 59(4):898–913, 2011. doi: 10.1287/opre.1110.0926.

[5] Dimitris Bertsimas and Melvyn Sim. The price of robustness. *Operations Research*, 52(1):35–53, 2004.

[6] John R. Birge. Option methods for incorporating risk into linear capacity planning models. *Manufacturing & Service Operations Management*, 2 (1):19–31, 2000.

[7] John R. Birge and François Louveaux. *Introduction to Stochastic Programming.* Springer, Berlin Heidelberg New York, 1997.

[8] George B. Dantzig and Gerd Infanger. Large-scale stochastic linear programs—importance sampling and Benders decomposition. In *Computational and applied mathematics, I (Dublin, 1991)*, pages 111–120. North-Holland, Amsterdam, 1992.

[9] Jitka Dupačová and Werner Römisch. Quantitative stability for scenario-based stochastic programs. In Marie Hušková, Petr Lachout, and Jan Ámos Víšek, editors, *Prague Stochastics '98*, pp 119–124. JČMF, 1998.

A.J. King and S.W. Wallace, *Modeling with Stochastic Programming*,
Springer Series in ORFE, DOI 10.1007/978-0-387-87817-1,
© Springer Science+Business Media New York 2012

[10] Jitka Dupačová, Nicole Gröwe-Kuska, and Werner Römisch. Scenario reduction in stochastic programming: An approach using probability metrics. *Mathematical Programming*, 95(3):493–511, 2003. doi: 10.1007/s10107-002-0331-0.

[11] M. Ehrgott and D.M. Ryan. Constructing robust crew schedules with bicriteria optimization. *Journal of Multi-Criteria Decision Analysis*, 11(3):139–150, 2002.

[12] Matthias Ehrgott and David M. Ryan. The method of elastic constraints for multiobjective combinatorial optimization and its application in airline crew scheduling. In T. Tanino, T. Tanaka, and M. Inuiguchi, editors, *Multi-Objective Programming and Goal Programming – Theory and Applications*, pages 117–122. Springer, Berlin Heidelberg New York, 2003.

[13] Y. Ermoliev. Stochastic quasigradient methods and their application to system optimization. *Stochastics*, 9:1–36, 1983.

[14] Olga Fiedler and Werner Römisch. Stability in multistage stochastic programming. *Annals of Operations Research*, 56(1):79–93, 2005. doi: 10.1007/BF02031701.

[15] K. Froot and J. Stein. Risk management, capital budgeting and capital structure policy for financial institutions: An integrated approach. *Journal of Financial Economics*, 47:55–82, 1998.

[16] A. Gaivoronski. Stochastic quasigradient methods and their implementation. In *Numerical techniques for stochastic optimization*, volume 10 of *Springer Ser. Comput. Math.*, pp 313–351. Springer, Berlin Heidelberg New York, 1988.

[17] V. Gaur and S. Seshadri. Hedging inventory risk through market instruments. *Manufacturing and Service Operations Management*, 7(2):103–120, 2005.

[18] R.C. Grinold. Model building techniques for the correction of end effects in multistage convex programs. *Operations Research*, 31(4):407–431, 1983.

[19] J. Michael Harrison and Stanley R. Pliska. Martingales and stochastic integrals in the theory of continuous time trading. *Stochastic Processes and Their Applications*, 11:215–260, 1981.

[20] H. Heitsch and W. Römisch. Scenario reduction algorithms in stochastic programming. *Computational Optimization and Applications*, 24(2–3):187–206, 2003. doi: 10.1023/A:1021805924152.

[21] H. Heitsch, W. Römisch, and C. Strugarek. Stability of multistage stochastic programs. *SIAM Journal on Optimization*, 17(2):511–525, 2006. doi: 10.1137/050632865.

[22] Holger Heitsch and Werner Römisch. A note on scenario reduction for two-stage stochastic programs. *Operations Research Letters*, 35(6):731–738, 2007. doi: 10.1016/j.orl.2006.12.008.

[23] Holger Heitsch and Werner Römisch. Scenario tree reduction for multistage stochastic programs. *Computational Management Science*, 6(2):117–133, 2009. doi: 10.1007/s10287-008-0087-y.

[24] J. L. Higle and S. Sen. Stochastic decomposition: An algorithm for two-stage linear programs with recourse. *Mathematics of Operations Research*, 16:650–669, 1991.

[25] J. L. Higle and S. Sen. Statistical verification of optimality conditions for stochastic programs with recourse. *Annals of Operations Research*, 30: 215–240, 1991.

[26] J. L. Higle and S. W. Wallace. Sensitivity analysis and uncertainty in linear programming. *Interfaces*, 33:53–60, 2003.

[27] K. Høyland and S. W. Wallace. Generating scenario trees for multistage decision problems. *Management Science*, 47(2):295–307, 2001. doi: 10. 1287/mnsc.47.2.295.9834.

[28] Kjetil Høyland, Michal Kaut, and Stein W. Wallace. A heuristic for moment-matching scenario generation. *Computational Optimization and Applications*, 24(2–3):169–185, 2003. ISSN 0926-6003.

[29] Gerd Infanger. Monte Carlo (importance) sampling within a Benders decomposition algorithm for stochastic linear programs. *Annals of Operations Research*, 39(1–4):69–95 (1993), 1992. ISSN 0254-5330.

[30] P. Kall and S.W. Wallace. *Stochastic Programming*. Wiley, Chichester, 1994.

[31] Michal Kaut and Stein W. Wallace. Evaluation of scenario-generation methods for stochastic programming. *Pacific Journal of Optimization*, 3 (2):257–271, 2007.

[32] Michal Kaut and Stein W. Wallace. Shape-based scenario generation using copulas. *Computational Management Science*, 8(1–2):181–199, 2011. doi: 10.1007/s10287-009-0110-y.

[33] Michal Kaut, Stein W. Wallace, Hercules Vladimirou, and Stavros Zenios. Stability analysis of portfolio management with conditional value-at-risk. *Quantitative Finance*, 7(4):397–409, 2007. doi: 10.1080/ 14697680701483222.

[34] A. J. King. Asymmetric risk measures and tracking models for portfolio optimization under uncertainty. *Annals of Operations Research*, 45: 165–177, 1993.

[35] Alan J. King. Duality and martingales: A stochastic programming perspective on contingent claims. *Mathematical Programming*, 91(3): 543–562, 2002.

[36] Alan J. King, Teemu Pennanen, and Matti Koivu. Calibrated option bounds. *International Journal of Theoretical and Applied Finance*, 8:141–159, 2005.

[37] Alan J. King, Olga Streltchenko, and Yelena Yesha. Private valuation of contingent claims in a discrete time/state model. In John B. Guerard, editor, *Handbook of Portfolio Construction: Contemporary Applications*, pp 691–710. Springer, Berlin Heidelberg New York, 2010.

[38] Anton J. Kleywegt, Alexander Shapiro, and Tito Homem-de Mello. The sample average approximation method for stochastic discrete optimization. *SIAM Journal on Optimization*, 12(2):479–502, 2001. doi: 10.1137/S1052623499363220.

[39] A. G. Kök, M.L. Fisher, and R. Vaidyanathan. Assortment planning: Review of literature and industry practice. In N. Agrawal and S.A. Smith, editors, *Retail Supply Chain Management*, pp 99–154. Springer, Berlin Heidelberg New York, 2008.

[40] A.-G. Lium, T. G. Crainic, and S. W. Wallace. A study of demand stochasticity in stochastic network design. *Transportation Science*, 43(2): 144–157, 2009. doi: 10.1287/trsc.1090.0265.

[41] Leonard C. MacLean, Edward O. Thorp, and William T. Ziemba. *The Kelly capital growth investment criterion: Theory and practice*. Handbook in Financial Economics. World Scientific, Singapore, 2011.

[42] S. Mahajan and G. van Ryzin. Retail inventories and consumer choice. In S. Tayur, R. Ganesham, and M. Magasine, editors, *Quantitative methods in Supply Chain Management*. Kluwer, Amsterdam, 1998.

[43] W.K. Mak, D.P. Morton, and R.K. Wood. Monte carlo bounding techniques for determining solution quality in stochastic programs. *Operations Research Letters*, 24:47–56, 1999.

[44] H.M. Markowitz. *Portfolio selection: Efficient diversification of investment*. Yale University Press, New Haven, CT, 1959.

[45] G. C. Pflug. Scenario tree generation for multiperiod financial optimization by optimal discretization. *Mathematical Programming*, 89(2): 251–271, 2001. doi: 10.1007/PL00011398.

[46] András Prékopa. *Stochastic programming*, vol 324. *Mathematics and Its Applications*. Kluwer, Dordrecht, 1995. ISBN 0-7923-3482-5.

[47] J. M. Rosenberger, E. L. Johnson, and G. L. Nemhauser. A robust fleet-assignment model with hub isolation and short cycles. *Transportation Science*, 38(3):357–368, 2004.

[48] Andrej Ruszczynski and Alexanger Shapiro. *Stochastic Programming*. Handbooks in Operations Research and Management Science. Elsevier, Amsterdam, 2002.

[49] Alexander Shapiro. Monte carlo sampling approach to stochastic programming. *ESAIM: Proceedings*, 13:65–73, 2003. doi: 10.1051/proc: 2003003. Proceedings of 2003 MODE-SMAI Conference.

[50] Alexander Shapiro. Monte Carlo sampling methods. In A. Ruszczyński and A. Shapiro, editors, *Stochastic Programming*, volume 10 of *Handbooks in Operations Research and Management Science*, chapter 6, pp 353–425. Elsevier, Amsterdam, 2003. doi: 10.1016/S0927-0507(03)10006-0.

[51] Gordon Sick. Real options. In *Finance*, vol 9. *Handbooks in Operations Research and Management Science*, chap 21, pp 631–691. Elsevier, Amsterdam, 1995.

[52] J.W. Suurballe and Robert E. Tarjan. A quick method for finding shortest pairs of paths. *Networks*, 14:325–336, 1984.

[53] Hajnalka Vaagen and Stein W. Wallace. Product variety arising from hedging in the fashion supply chains. *International Journal of Production Economics*, 114(2):431–455, 2008. doi: 10.1016/j.ijpe.2007.11.013.

[54] J. von Neumann and O. Morgenstern. *Theory of Games and Economic Behavior*, 2nd edn. Princeton University Press, Princeton, NJ, 1947.

[55] S.W. Wallace. Decision making under uncertainty: Is sensitivity analysis of any use? *Operations Research*, 48:20–25, 2000.

[56] S.W. Wallace and W.T. Ziemba, editors. *Applications of Stochastic Programming*. MPS-SIAM Series on Optimization, Philadelphia, 2005.

Index

A
approximation error, 101
Arrow–Debreu pricing, 144
assortment planning, 125

B
bias, 86, 92
bimodal distribution, 127, 130

C
calibration, 143, 146
    liquidity constraints, 146
    real options model, 150
chance constrained model, **27**
chance-constrained model, 37, 54
Cholesky decomposition, 96
complicated distributions, 123
conditional value at risk, *see* CVaR
consolidation, 104, 117, 118
copula, 100
corporate risk-taking, 64
correlations, 22
CVaR, 73, 74

D
dependent random variables, 21
discount rate, 139
discounting, 55
distribution
    bimodal, 127, 130
    correlations, 111–114, 121, 127, 133
    dependence, 21, 47
    existence, 18
    independent, 100

mis-specifying, 133
    partial information, 20
    uncorrelated, 100
    varance-covariance matrix, 71
distribution of outcomes, 61
dual equilibrium, 56
dynamics, 33

E
efficient frontier, 69
    Markowitz, 72
empirical distribution, 81
event tree, **6**
example
    airline, 26
    electricity production, 16, 23
    fashion, 123–138
    inventory model, 49
        chance-constrained, 54
        worst-case, 53
    knapsack problem, 33
        chance-constrained, 37
        multistage, 39
        objective function, 62
        stochastic robust, 38
        two-stage model, 35, 36
        worst-case, 36, 37
    long lead production, 153–164
    network design, 86
    newsmix problem, 4
    oil platform, 24
    oil-well abandonment, 25
    options pricing, 140

A.J. King and S.W. Wallace, *Modeling with Stochastic Programming*,
Springer Series in ORFE, DOI 10.1007/978-0-387-87817-1,
© Springer Science+Business Media New York 2012

overhaul project, 39
  inherently two-stage, 45
  two-stage model, 43
  worst-case, 46
risk management, 21, 24
service network design, 103–122
sports event, 13, 23
telecommunications, 27
truck routing, 12, 23
expected objective function value, 63,
    65, 66
expected utility, 69, 70
extreme event, 73
  different distributions for, 75

F
fashion, 123–138, 162
feasibility, 5, **33**, 59
financial markets, 139
flexibility, **12**, 14, 111, 112, 115, 116, 119,
    120
  bounding using . . . , 14
forecasting demand, 157

H
hedging, 118
horizon effect, 55, 108

I
implementability, 52
implicity option, 116
in-sample stability, 83, 84, 87, 92, 106,
    111
information stage, *see* stage
information structure, 50
inherently multistage models, 16
inherently two-stage models, **16**, 34, 45,
    107, 110, 123
inventory, 157
invest-and-use models, *see* inherently
    two-stage models

K
knowledge about the future, 19

L
learning, 76, 155
less-than-truckload trucking, 103, 104
luck, 76

M
markets, 139
Markowitz
  approximate utility, 71
Markowitz model, 70, 81
martingale, 143, 144
mis-specifying distributions, 133
moment matching methods, *see*
    property matching methods
multiobjective, 69
multiple risks, 69
multistage formulation, 39

N
negative correlations, 121
net present value, 25
news mix problem
  sensitivity analysis, 5
newsboy, 123, 148
newsmix problem, 4
  two-stage formulation, 8
nonanticipativity, 52

O
objective function, 13, **61**
  expected value, 63, 65, 66
  option, 66, 68
  penalty, 66
  recourse, 66, 68
  shortfall, 66
  target, 66
operating cost, 26
operational models, *see* inherently
    multistage models
operational risk, 75
optimality gap, 88, 90, 92, 101
option, 66, 68
options
  relation to recourse, 115
options pricing
  errors, 151
  linear programming duality, 142, 143
  replication argument, 141, 145
  risk neutrality, 144
  simple, 140
options theory, 26
out-of-sample stability, 83, 85, 87, 92,
    111
  multiperiod trees, 87

P

penalty, 66
portfolio, 135
property matching methods, 92
   regression models, 94
   software, 99
   transformation model, 95–100

Q

quality of solution, 88
   optimality gap, 88, 90, 92
   statistical estimate, 88
   stochastic upper bound, 89

R

real options, 25
real options model, 150, 163
real options theory, 24
recourse option, 26, 27, 66, 68, 116
recovery cost, 26
rejected demand, 104
replication argument, 141, 145
risk-neutral pricing, 144
robust optimization, 28
robustness, **12**, 14, 111
   bounding using . . . , 15

S

sample average approximation, 90, 102
sampling, 79
scenario analysis, 2, 3
scenario generation, 20, 77–102, 106
   multi-modal distributions, 127
   property matching, 92
   quality of . . . , 78, 81
   sampling, 79
   stability, 83
sensitivity analysis, 2, 3, 5, 10
   deterministic models, 11, 49

share-holder, 64
shortfall, 66, 161
soft constraint, 20, 62, 158, 160
   rejected demand, 104
space-time network, 108
stability testing, 83, 88, 91
stage, 6, 9, 107
   electricity production model, 16
   long lead time production, 154
statistical approaches to solution
   quality, 88
steady-state, 15, 31
stochastic discount factors, **139**, 145
   calibration, 146
   options, 144
stochastic dynamic programming, 31
stochastic robust optimization, 29, 38
   feasibility, 30
   interval sensitivity, 30
stress-test, 2, 3
substitution, 127, 131
supplier-managed inventory, 153

T

tail behaviour, 73
target, 66
transient modeling, 15

V

value at risk, *see* VaR
VaR, 73, 74
variance reduction, 90

W

Wasserstein metric, 101
what-if analysis, 2, 3, 10
   deterministic models, 11
worst-case, 20, 36, 37, 46, 53